Cohomological Methods in Group Theory

PURE AND APPLIED MATHEMATICS

A Series of Monographs and Textbooks

1. K. YANO. Integral Formulas in Riemannian Geometry (1970)
2. S. KOBAYASHI. Hyperbolic Manifolds and Holomorphic Mappings (1970)
3. V. S. VLADIMIROV. Equations of Mathematical Physics (A. Jeffrey, editor; A. Littlewood, translator) (1970)
4. B. N. PSHENICHNYI. Necessary Conditions for an Extremum (L. Neustadt, translation editor; K. Makowski, translator) (1971)
5. L. NARICI, E. BECKENSTEIN, and G. BACHMAN. Functional Analysis and Valuation Theory (1971)
6. D. S. PASSMAN. Infinite Group Rings (1971)
7. L. DORNHOFF. Group Representation Theory (in two parts). Part A: Ordinary Representation Theory. Part B: Modular Representation Theory (1971, 1972)
8. W. BOOTHBY and G. L. WEISS (eds.). Symmetric Spaces: Short Courses Presented at Washington University (1972)
9. Y. MATSUSHIMA. Differentiable Manifolds (E. T. Kobayashi, translator) (1972)
10. L. E. WARD, JR. Topology: An Outline for a First Course (1972)
11. A. BABAKHANIAN. Cohomological Methods in Group Theory (1972)

In Preparation:

R. GILMER. Multiplicative Ideal Theory

J. M. MENDEL. Discrete Techniques of Parameter Estimation: The Equation Error Formulation

Cohomological Methods in Group Theory

ARARAT BABAKHANIAN

DEPARTMENT OF MATHEMATICS
UNIVERSITY OF ILLINOIS
URBANA, ILLINOIS 61801

1972

MARCEL DEKKER, INC., New York

MARCEL DEKKER, INC.
95 Madison Avenue, New York, New York 10016

LIBRARY OF CONGRESS CATALOG CARD NUMBER: 70–182212

ISBN: 0–8247–1031–2

PRINTED IN THE UNITED STATES OF AMERICA

Preface

This book is intended for students interested in learning the use of cohomology and homology theory in solving problems in group theory. Although cohomology groups of a group were formally defined in the early 1940's, these groups in low dimensions had been studied earlier as part of the general body of theory of groups. For instance, for finite groups, group extensions, the Schur multiplicator, a group modulo its commutator subgroup are directly expressible as certain cohomology groups.

In the last three decades cohomology of groups has played a central role in various branches of mathematics. This theory was first formally studied by Eilenberg and MacLane in connection with their investigations in algebraic topology. Aside from their work, the works of Cartan and Leray on spectral sequences and the use of these sequences by Roger Lyndon and later by Hochschild and Serre in the forties and the early fifties in cohomology theory of group extensions revealed some of the basic structure in the theory. The theory flourished in the fifties because of its application by Artin and Tate to class field theory. Many basic properties of cohomology of a group and its comparison with the cohomology groups of its subgroups were discovered, notably by Artin, Tate, Chevalley, Eckmann, Hochschild, and Nakayama. Golod, Venkov, and Evens discovered the finite generation of the cohomology ring of finite groups with suitable coefficient groups. The discovery of these properties gave new impetus to the already vast investigation of cohomology theory of groups. Successful attempts were made to answer some purely group theoretic questions with the help of cohomology theory. Examples of such works are Tate's result on the existence of normal p-complement using the Hochschild-Serre five-term exact sequence and Stallings' use of Tate's method in the study of descending central series. This gave as corollaries some of the well-known theorems of Magnus. Cohomological triviality was used by Gaschütz in his proof of the existence of p-automorphisms of nonabelian p-groups.

Other interesting results in cohomology of groups are Stallings' proof of a conjecture by Eilenberg and Ganea that a finitely generated group of cohomological dimension 1 is free, and his proof of a conjecture of Serre that a torsion-free, finitely generated group having a free subgroup of finite index is itself free. Another recent result is Quillen's proof of a conjecture by Atiyah and Swan that the Krull dimension of the mod p cohomology ring of a finite group equals the maximum rank of an elementary abelian p-subgroup.

Our attempt here is to provide the reader with the basic tools in cohomology of groups and to illustrate their use in obtaining group theoretic results. The material included grew out of my lecture notes for two one-semester courses at the University of Illinois in the fall of 1967 and the spring of 1970. The first of these lecture notes appeared as No. 17 in Queen's Papers in Pure and Applied Mathematics. Some of the chapters can be used for a one-semester course. For instance, Chapters 1, 2, 3, 5, 7, 10, or, 1, 2, 3, 5, 6, 8, 9 are suitable for such purposes. The bibliography is not a complete listing of all the works on and related to cohomology of groups, but is intended as a guide to further reading. References are made by sections. For instance, the reference 36.7 means subsection 7 of section 36. Square brackets refer to the bibliography.

Urbana, Illinois
March, 1972

ARARAT BABAKHANIAN

Contents

Cohomological Methods in Group Theory

CHAPTER

1

Elements of Homological Algebra

This chapter contains the basic algebraic tools needed in cohomology theory of groups.

1. Modules

1.1 Let R be a ring with identity. A *free R-module* is a direct sum, $\sum_{\iota \in I} R_\iota$, $R_\iota = R$ for every $\iota \in I$, of copies of R. For convenience we will denote the free R-module $\sum_{\iota \in I} R_\iota$ by $\sum_{\iota \in I} Rx_\iota$ where x_ι are distinct symbols. Thus the free R-module consists of elements of the form

$$\alpha_{\iota_1} x_{\iota_1} + \cdots + \alpha_{\iota_s} x_{\iota_s} \qquad \text{with} \quad \alpha_{\iota_j} \in R$$

where $\sum_{j=1}^{s} \alpha_{\iota_j} x_{\iota_j} = 0$ will mean $\alpha_{\iota_j} = 0$ for $j = 1, \ldots, s$.

1.2 Let $A = \sum_{\iota \in I} Rx_\iota$ be a free R-module, B an R-module. Then

(i) an R-homomorphism $f\colon A \to B$ is completely determined if the images $f(1 \cdot x_\iota)$ are given;

(ii) if $\{b_\iota\}_{\iota \in I}$ is an arbitrary family of elements of B indexed by the set I, then there exists an R-homomorphism $f: A \to B$ such that $f(1x_\iota) = b_\iota$.

1.3 *Every R-module is a factor module of a free R-module.*

PROOF Let A be an R-module. Let $\{a_\iota\}_{\iota \in I}$ be any set of generators of the R-module A, e.g., all the elements of A. Let $F = \sum_{\iota \in I} Rx_\iota$ be a free R-module freely generated by the symbols x_ι, $\iota \in I$. Define the R-homomorphism $f: F \to A$ by $f(1x_\iota) = a_\iota$; f is an epimorphism since $f(\sum \alpha_\iota x_\iota) = \sum \alpha_\iota a_\iota$. If K is the kernel of f, we have $A \cong F/K$.

1.4 Let A and B be R-modules. If

$$f: A \to B \qquad \text{and} \qquad g: A \to B$$

are R-homomorphisms, define the R-homomorphism $f + g: A \to B$ by $(f + g)(a) = f(a) + g(a)$; $f + g$ is a homomorphism for,

$$\begin{aligned}(f + g)(a_1 + a_2) &= f(a_1 + a_2) + g(a_1 + a_2)\\ &= f(a_1) + f(a_2) + g(a_1) + g(a_2)\\ &= (f + g)(a_1) + (f + g)(a_2).\end{aligned}$$

$f + g$ is an R-homomorphism for,

$$\begin{aligned}(f + g)(ra) &= f(ra) + g(ra)\\ &= rf(a) + rg(a)\\ &= r(f + g)(a).\end{aligned}$$

Similarly, $f - g: A \to B$ defined by $(f - g)(a) = f(a) - g(a)$ is an R-homomorphism. Thus $\mathrm{Hom}_R(A, B)$, the set of R-homomorphisms from A to B, has the structure of an abelian group. In general $\mathrm{Hom}_R(A, B)$ is not an R-module. For, we must have $rf(r'a)$ equal to $r' \cdot rf(a)$ and also equal to $r \cdot r'f(a)$. So $\mathrm{Hom}_R(A, B)$ has the structure of an R-module if R is commutative.

1.5. Given the R-homomorphisms $f: A \to B$ and $g: B \to C$, $gf: A \to C$ is an R-homomorphism defined by $(gf)(a) = g(f(a))$. (We remark that the

R-modules and the R-homomorphisms form a category. This category is abelian.)

2. Tensor Products

In this section R is a *commutative* ring with identity.

2.1 Let A, B, and T be left R-modules. A map $\theta: A \times B \to T$ is called a *bilinear map* if

$$\theta(a_1 + a_2, b) = \theta(a_1, b) + \theta(a_2, b),$$
$$\theta(a, b_1 + b_2) = \theta(a, b_1) + \theta(a, b_2),$$
$$\theta(ra, b) = \theta(a, rb) = r\theta(a, b).$$

We say the R-module T together with the bilinear map $\theta: A \times B \to T$ represents the *tensor product* of A and B if the following conditions are satisfied:

(T1) The image of $A \times B$ under θ generates T.

(T2) If C is an R-module and $f: A \times B \to C$

is a bilinear map, then there exists an R-homomorphism $\bar{f}: T \to C$ such that the diagram

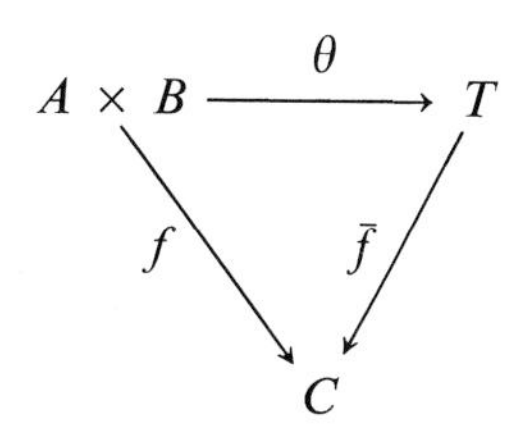

is commutative.

We note that since $\bar{f}(\theta(a, b)) = f(a, b)$ for every generator $\theta(a, b)$ of T, $\bar{f}$ is uniquely determined by f.

Assume T_1, θ_1 and T_2, θ_2 represent the tensor product of the R-modules A and B. Then there exist unique R-homomorphisms $\bar{\theta}_1: T_2 \to T_1$ and $\bar{\theta}_2: T_1 \to T_2$ such that

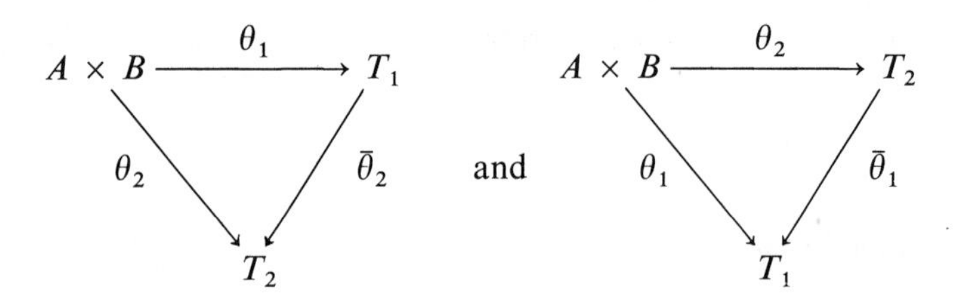

are commutative. Thus $\theta_2 = \bar{\theta}_2\bar{\theta}_1\theta_2$; i.e., $\bar{\theta}_2\bar{\theta}_1$ leaves a set of generators of T_2 fixed. Hence $\bar{\theta}_2\bar{\theta}_1$ is identity on T_2. Similarly $\bar{\theta}_1\bar{\theta}_2$ leaves T_1 fixed. Therefore $\bar{\theta}_2: T_1 \to T_2$ is an isomorphism. So the tensor product of A and B is unique up to a canonical isomorphism. For this reason we will denote any representative T, θ of the tensor product of the R-modules A and B by $A \times B \xrightarrow{\theta} A \otimes_R B$, or simply by $A \otimes_R B$.

2.2 Assuming the tensor product $A \otimes_R B$ exists we write $\theta(a, b) = a \otimes b$. The bilinearity of θ in this notation means

$$(a_1 + a_2) \otimes b = a_1 \otimes b + a_2 \otimes b,$$

$$a \otimes (b_1 + b_2) = a \otimes b_1 + a \otimes b_2$$

$$r(a \otimes b) = ra \otimes b = a \otimes rb.$$

$A \otimes_R B$ is generated as an R-module by the elements $a \otimes b$; i.e., an element of $A \otimes_R B$ has the form $\sum_\iota r_\iota(a'_\iota \otimes b'_\iota) = \sum_\iota a_\iota \otimes b_\iota$. This representation of the elements of $A \otimes_R B$ is not unique.

3. Existence of Tensor Products

In this section R is a commutative ring with identity.

3.1 *The R-module A represents the tensor product $R \otimes_R A$.*

PROOF Define $\theta: R \times A \to A$ by $\theta(r, a) = ra$. Axiom (T1) is satisfied since A is generated by all ra. Axiom (T2) is satisfied for, if $f: R \times A \to C$ is bilinear, then define $\bar{f}: A \to C$ by $\bar{f}(ra) = f(r, a)$; $\bar{f}$ is well defined since f is bilinear. Thus the R-module A represents $R \otimes_R A$.

3.2 *Let $A = \sum_{\iota \in I} A_\iota$ be a direct sum of R-modules. Suppose for every $\iota \in I$ and an R-module B the tensor products $A_\iota \otimes_R B$ exist. Then $\sum_{\iota \in I} (A_\iota \otimes_R B)$ represents the tensor product $A \otimes_R B$.*

PROOF Define $\theta: A \times B \to \sum_{\iota \in I} (A_\iota \otimes_R B)$ by $\theta(\sum a_\iota, b) = \sum (a_\iota \otimes b)$. The elements $a_\iota \otimes b$ generate $A_\iota \otimes_R B$ for every $\iota \in I$. Therefore Axiom (T1) is satisfied. Let $f: A \times B \to C$ be bilinear. Let $f_\iota: A_\iota \times B \to C$ be

$$A_\iota \times B \xrightarrow{i_\iota \times 1_B} A \times B \xrightarrow{f} C$$

where i_ι is the injection of the summand A_ι into the direct sum $A = \sum_\iota A_\iota$; f_ι is bilinear, hence there exists $\bar{f}_\iota: A_\iota \otimes B \to C$ such that

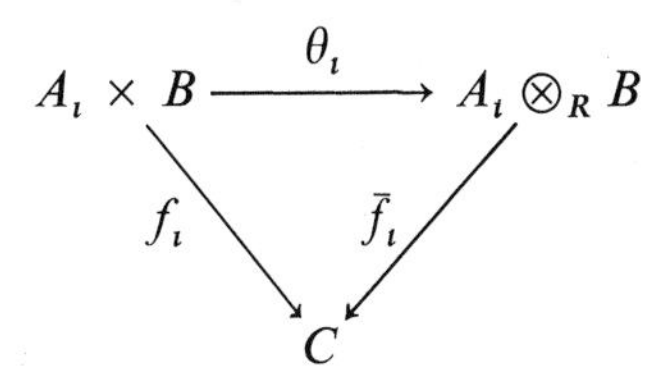

is commutative. Define $\bar{f}: \sum_\iota (A_\iota \otimes_R B) \to C$ by

$$\bar{f}\left(\sum_{\iota,\kappa} a_{\iota\kappa} \otimes b_{\iota\kappa}\right) = \sum_\iota \bar{f}_\iota \left(\sum_\kappa a_{\iota\kappa} \otimes b_{\iota\kappa}\right).$$

The verification that $\bar{f}$ is an R-homomorphism is straightforward. Axiom (T2) is satisfied since

$$\begin{aligned} \bar{f}\theta\left(\sum a_\iota, b\right) &= \bar{f}\left(\sum (a_\iota \otimes b)\right) \\ &= \sum \bar{f}_\iota (a_\iota \otimes b) \\ &= \sum f_\iota (a_\iota, b) \\ &= f\left(\sum a_\iota, b\right). \end{aligned}$$

3.3 It follows from 3.1 and 3.2 that:

If A is a free R-module, and B is any R-module, then $A \otimes_R B$ exists and is represented by a direct sum of R-modules isomorphic to B.

3.4 *Let M be an R-submodule of A and N an R-submodule of B. Suppose $A \otimes_R B$ exists. Let E be the submodule of $A \otimes_R B$ generated by all $m \otimes b$*

and $a \otimes n$, where $m \in M$, $a \in A$, $n \in N$, and $b \in B$. Then the tensor product $A/M \otimes_R B/N$ exists and is represented by $A \otimes_R B/E$.

PROOF Define $\theta\colon A/M \times B/N \to A \otimes_R B/E$ by $\theta(a + M, b + N) = a \otimes b + E$; θ is well defined for if $m \in M$ and $n \in N$, then

$$\begin{aligned}\theta(a + m + M, b + n + N) &= a \otimes b + a \otimes n + m \otimes b + m \otimes n + E \\ &= a \otimes b + E\end{aligned}$$

The verification that θ is bilinear is straightforward and is left to the reader. Axiom (T1) is satisfied since the elements $a \otimes b + E = \theta(a + M, b + N)$ generate $A \otimes_R B/E$.

Let $f\colon A/M \times B/N \to C$ be a bilinear map. Define $f_1\colon A \times B \to C$ by $f_1(a, b) = f(a + M, b + N)$; f_1 is well defined and bilinear since f_1 is the composite map

$$A \times B \xrightarrow{c} A/M \times B/N \xrightarrow{f} C \qquad \text{where} \qquad c(a, b) = (a + M, b + N),$$

i.e., f_1 is an R-homomorphism followed by a bilinear map. Hence there exists $\bar{f}_1\colon A \otimes B \to C$ such that $f_1(a, b) = \bar{f}_1(a \otimes b)$. For $m \in M$, $b \in B$,

$$\bar{f}_1(m \otimes b) = f_1(m, b) = f(M, b + N) = 0.$$

Similarly for $a \in A$, $n \in N$, $\bar{f}_1(a \otimes n) = 0$. Therefore E is in the kernel of $\bar{f}_1$. Define $\bar{f}\colon A \otimes_R B/E \to C$ by $\bar{f}(\sum a \otimes b + E) = \bar{f}_1(\sum a \otimes b)$; $\bar{f}$ is well defined for if $e \in E$, then

$$\begin{aligned}\bar{f}(\textstyle\sum a \otimes b + e + E) &= \bar{f}_1(\textstyle\sum a \otimes b + e) \\ &= \bar{f}_1(\textstyle\sum a \otimes b) + 0 \\ &= \bar{f}(\textstyle\sum a \otimes b + E).\end{aligned}$$

Moreover

$$f(a + M, b + N) = f_1(a, b) = \bar{f}_1(a \otimes b) = \bar{f}(a \otimes b + E).$$

Therefore Axiom (T2) is satisfied.

3.5 *The tensor product of any two R-modules exists.*

PROOF Let A, B be R-modules. By 1.3, $A \cong F/K$ where F and K are

R-modules and F is a free module; $F \otimes_R B$ exists by 3.3 and $(F/R) \otimes_R B$ exists by 3.4. The isomorphism $A \otimes_R B \cong (F/R) \otimes_R B$ establishes the existence of the tensor product of two R-modules.

3.6. Let A and B be R-modules. We can construct, canonically, a representation of the tensor product $A \otimes_R B$ as follows: Let $F_{A \times B}$ be the free R-module generated by the elements of $A \times B$, then the factor module defined by the relations

$$(a_1 + a_2, b) - (a_1, b) - (a_2, b) \qquad (ra, b) - (a, rb)$$

$$(a, b_1 + b_2) - (a, b_1) - (a, b_2) \qquad r(a, b) - (ra, b)$$

in $F_{A \times B}$ represents $A \otimes_R B$.

We leave the verification that Axioms (T1) and (T2) are satisfied to the reader.

3.7 Consider the two **Z**-modules $\mathbf{Z}/(2)$ and $\mathbf{Z}/(3)$. where **Z** is the ring of integers; $\mathbf{Z}/(2)$ has elements 0, x and $\mathbf{Z}/(3)$ has elements 0, y, $2y$. Observe that

$$x \otimes y = 3x \otimes y = x \otimes 3y = x \otimes 0 = 0$$

$$x \otimes 2y = 2x \otimes y = 0 \otimes y = 0$$

Therefore $\mathbf{Z}/(2) \otimes_{\mathbf{Z}} \mathbf{Z}/(3) = 0$.

4. The Homomorphism $f \otimes g$

In this section R is a commutative ring with identity. We shall write $A \otimes B$ instead of $A \otimes_R B$ when it is clear that A and B are R-modules.

4.1 Let $f: A \to A'$, $g: B \to B'$ be R-homomorphisms. Consider the map $F: A \times B \to A' \otimes B'$ defined from the underlying set of $A \times B$ to the underlying set of $A' \otimes B'$ by $F(a, b) = f(a) \otimes g(b)$; F is bilinear. For,

$$\begin{aligned} F(a_1 + a_2, b) &= f(a_1 + a_2) \otimes g(b) \\ &= f(a_1) \otimes g(b) + f(a_2) \otimes g(b) \\ &= F(a_1, b) + F(a_2, b). \end{aligned}$$

Similarly

$$F(a, b_1 + b_2) = F(a, b_1) + F(a, b_2)$$

and

$$F(ra, b) = F(a, rb) = rF(a, b).$$

By the definition of tensor product, there exists an R-homomorphism $\bar{F}: A \otimes B \to A' \otimes B'$ defined on the generators of $A \otimes B$ by $\bar{F}(a \otimes b) = F(a, b)$. So $\bar{F}(\sum a \otimes b) = \sum f(a) \otimes g(b)$. We will denote $\bar{F}$ by $f \otimes g$.

4.2 Consider the R-homomorphisms

$$A \xrightarrow{f} A' \xrightarrow{f'} A'', \qquad B \xrightarrow{g} B' \xrightarrow{g'} B''.$$

Then we have two R-homomorphisms

$$f'f \otimes g'g: A \otimes B \to A'' \otimes B''$$

and

$$(f' \otimes g')(f \otimes g): A \otimes B \to A'' \otimes B''.$$

Applying the latter R-homomorphism to a generator $a \otimes b$ of $A \otimes B$ we get

$$\begin{aligned} (f' \otimes g')(f \otimes g)(a \otimes b) &= (f' \otimes g')(f(a) \otimes g(b)) \\ &= f'f(a) \otimes g'g(b) \\ &= (f'f \otimes g'g)(a \otimes b). \end{aligned}$$

Thus

$$(f' \otimes g')(f \otimes g) = f'f \otimes g'g.$$

4.3 If $f_1, f_2: A \rightrightarrows A'$ and $g: B \to B'$ are R-homomorphisms, then

$$(f_1 + f_2) \otimes g = f_1 \otimes g + f_2 \otimes g.$$

This identity follows from the definitions.

4.4 $A \otimes_R B \cong B \otimes_R A.$

Define $\tau: B \times A \to A \otimes_R B$ by $\tau(b, a) = a \otimes b$; τ is bilinear, hence there exists an R-homomorphism $\bar{\tau}: B \otimes_R A \to A \otimes_R B$ such that $\bar{\tau}(b \otimes a) = a \otimes b$. Similarly there exists $\bar{\sigma}: A \otimes_R B \to B \otimes_R A$ defined by $\bar{\sigma}(a \otimes b) = b \otimes a$. It follows from the uniqueness of $\bar{\sigma}$ and $\bar{\tau}$ that $\bar{\sigma}\bar{\tau} = 1_{B \otimes_R A}$ and $\bar{\tau}\bar{\sigma} = 1_{A \otimes_R B}$.

5. Tensor Product of Algebras

5.1. Let R be a commutative ring with identity, then the ring A (not necessarily with identity) is called an *R-algebra* if the following conditions are satisfied:

(i) A is a left R-module.

(ii) If $r_1, r_2 \in R$, $a_1, a_2 \in A$, then $r_1(a_1) \cdot r_2(a_2) = r_1 r_2(a_1 a_2)$.

5.2 Let A and B be R-algebras. We will impose an R-algebra structure on the R-module $A \otimes_R B$.

Define a multiplication in $A \otimes_R B$ by

$$\left(\sum a_\iota \otimes b_\iota\right)\left(\sum a'_\kappa \otimes b'_\kappa\right) = \sum a_\iota a'_\kappa \otimes b_\iota b'_\kappa$$

This multiplication is well defined. In fact, let $(a_0, b_0) \in A \times B$. The map $f_{(a_0,b_0)}: A \times B \to A \otimes_R B$ defined by $f_{(a_0,b_0)}(a, b) = aa_0 \otimes bb_0$ is bilinear. Therefore there exists a unique R-homomorphism $\bar{f}_{(a_0,b_0)}: A \otimes_R B \to A \otimes_R B$ such that the diagram

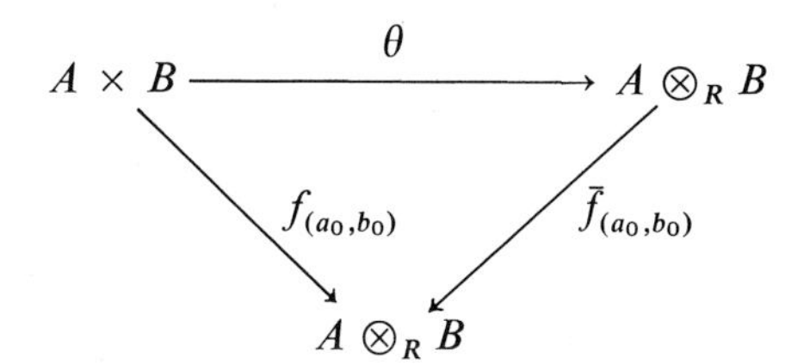

is commutative; i.e.,

$$\bar{f}_{(a_0,b_0)}(a \otimes b) = aa_0 \otimes bb_0.$$

Thus the multiplication of two elements of $A \otimes_R B$ is well defined if the second factor is a generator $a_0 \otimes b_0$ of $A \otimes_R B$. Since $\bar{f}_{(a,b)}$ is linear in a and in b, and

$$\bar{f}_{(ra_0,b_0)} = \bar{f}_{(a_0,rb_0)} = r\bar{f}_{(a_0,b_0)}$$

on any element $\sum a_\iota \otimes b_\iota$ of $A \otimes_R B$, the map

$$\bar{f}_{(\ ,\ )}(\textstyle\sum a_\iota \otimes b_\iota)\colon A \times B \to A \otimes_R B$$

is bilinear. Therefore there exists a unique R-homomorphism

$$\bar{f}_{(\ )}(\textstyle\sum a_\iota \otimes b_\iota)\colon A \otimes_R B \to A \otimes_R B$$

such that the diagram

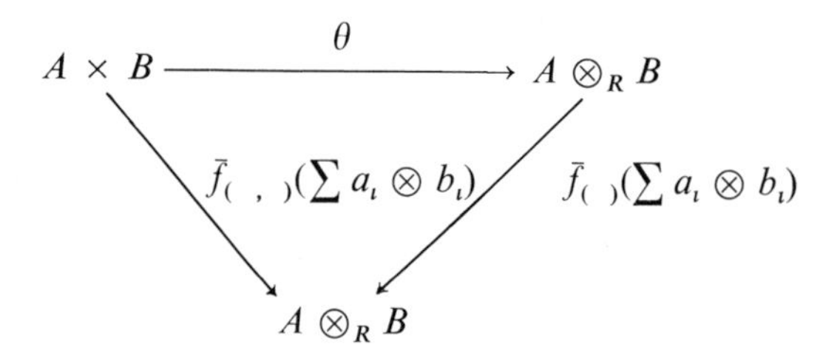

is commutative, i.e.,

$$\bar{f}_{(a' \otimes b')}(\textstyle\sum a_\iota \otimes b_\iota) = \sum a_\iota a' \otimes b_\iota b'.$$

Thus

$$\bar{f}_{(\sum_\kappa a_\kappa' \otimes b_\kappa')}(\sum_\iota a_\iota \otimes b_\iota) = \sum_\kappa \sum_\iota a_\iota a_\kappa' \otimes b_\iota b_\kappa'.$$

We have shown the multiplication

$$(\sum_\iota a_\iota \otimes b_\iota)(\sum_\kappa a_\kappa' \otimes b_\kappa') = \sum_{\iota,\kappa} a_\iota a_\kappa' \otimes b_\iota b_\kappa'$$

is well defined. The verification that $A \otimes_R B$ together with this multiplication is an R-algebra is left to the reader.

5.3 Let A, B, C be R-algebras. Let $f\colon A \to C$, $g\colon B \to C$ be algebra homomorphisms. Suppose $f(a)$ commutes with $g(b)$ for every $a \in A$ and $b \in B$. Then the map $F: A \times B \to C$ defined by $F(a, b) = f(a)g(b)$ is bilinear and the induced map $\bar{F}: A \otimes B \to C$ is an R-algebra homomorphism. Thus if A and B are commutative R-algebras with identity, the R-algebra $A \otimes_R B$ represents the direct sum of A and B in the category of commutative R-algebras. This follows from the axioms for tensor products. The injection maps $i_A: A \to A \otimes_R B$ and $i_B: B \to A \otimes_R B$ into the direct sum $A \otimes_R B$ are given by $i_A a = a \otimes 1$ and $i_B b = 1 \otimes b$. In particular the direct sum of two commutative rings A and B with identity is $A \otimes_Z B$.

6. Trilinear Maps

In this section R is a commutative ring with identity and A_1, A_2, A_3, B are R-modules.

6.1 The map $f: A_1 \times A_2 \times A_3 \to B$ is *trilinear* if

(i) for fixed $a_\iota \in A_\iota$, $a_\kappa \in A_\kappa$, f is an R-homomorphism of A_λ into B where ι, κ, λ are distinct elements of the set $\{1, 2, 3\}$;

(ii) $f(ra_1, a_2, a_3) = f(a_1, ra_2, a_3) = f(a_1, a_2, ra_3)$ for every $r \in R$.

We say the R-module T represents the tensor product $A_1 \otimes A_2 \otimes A_3$ if T is universal with respect to trilinear maps issued from $A_1 \times A_2 \times A_3$. The formal definition is similar to 2.1 where bilinear maps are replaced by trilinear maps.

6.2 *The tensor product $A_1 \otimes_R A_2 \otimes_R A_3$ exists and is represented by $(A_1 \otimes_R A_2) \otimes_R A_3$ [or $A_1 \otimes_R (A_2 \otimes_R A_3)$].*

Proof Let $f: A_1 \times A_2 \times A_3 \to B$ be a trilinear map. For a fixed $a_3 \in A_3$ define $f_{a_3}(a_1, a_2) = f(a_1, a_2, a_3)$, then $f_{a_3}: A_1 \times A_2 \to B$ is bilinear. Thus there exists an R-homomorphism $\bar{f}_{a_3}: A_1 \otimes_R A_2 \to B$ such that

$$\bar{f}_{a_3}(a_1 \otimes a_2) = f_{a_3}(a_1, a_2) = f(a_1, a_2, a_3).$$

Define the map $g: (A_1 \otimes_R A_2) \times A_3 \to B$ by

$$g\Big(\sum_{(a_1,a_2)} a_1 \otimes a_2, a_3\Big) = \bar{f}_{a_3}\Big(\sum_{(a_1,a_2)} a_1 \otimes a_2\Big).$$

g is bilinear. Hence there exists an R-homomorphism $\bar{g}: (A_1 \otimes_R A_2) \otimes_R A_3 \to B$ such that

$$\bar{g}((a_1 \otimes a_2) \otimes a_3) = f(a_1, a_2, a_3).$$

Let

$$\theta: A_1 \times A_2 \times A_3 \to (A_1 \otimes_R A_2) \otimes_R A_3$$

be the trilinear map defined by $\theta(a_1, a_2, a_3) = (a_1 \otimes a_2) \otimes a_3$; then the image of θ clearly generates $(A_1 \otimes_R A_2) \otimes_R A_3$. Moreover the diagram

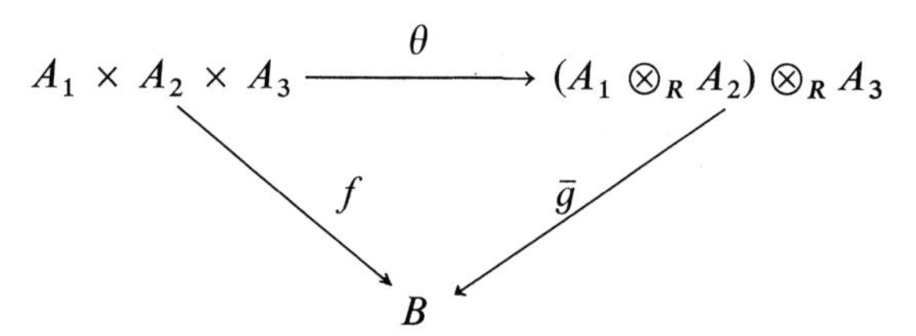

is commutative.

7. Exactness

Unless otherwise stated, the ring R is not necessarily commutative; A, B, C, A', B', . . . , are left (or right) R-modules; f, g, i, j, . . . , are R-homomorphisms.

7.1. The sequence $A \xrightarrow{f} B \xrightarrow{g} C$ is called *exact* at B if the kernel of g equals the image of f [i.e., $f(A) = g^{-1}(0)$].

EXAMPLES

(i) $0 \to A \xrightarrow{f} B$ is exact $\Leftrightarrow f$ is a monomorphism.

(ii) $B \xrightarrow{g} C \to 0$ is exact $\Leftrightarrow g$ is an epimorphism.

(iii) $0 \to A \xrightarrow{f} B \to 0$ is exact $\Leftrightarrow f$ is an isomorphism.

(iv) $0 \to A \to 0$ is exact $\Leftrightarrow A = 0$.

(v) $0 \to A \xrightarrow{i} B \xrightarrow{j} C \to 0$ is exact $\Leftrightarrow i$ is a monomorphism, j is an epimorphism, the image of i equals the kernel of j.

(vi) Let A be an R-submodule of the R-module B, then $0 \to A \xrightarrow{i} B \xrightarrow{j} B/A \to 0$ is exact where i is the inclusion of A into B and j is the induced canonical epimorphism.

7.2 *Let R be a commutative ring. If $0 \to A' \xrightarrow{f} A \xrightarrow{g} A'' \to 0$ is an exact sequence of R-homomorphisms and X is an R-module, then the sequence*

$$A' \otimes X \xrightarrow{f \otimes 1_X} A \otimes X \xrightarrow{g \otimes 1_X} A'' \otimes X \longrightarrow 0$$

is exact.

Proof If A' is an R-submodule of A and B' is an R-submodule of B, 3.4 restated in terms of exactness asserts that the sequence

$$(1) \qquad (A' \otimes B) \oplus (A \otimes B') \xrightarrow{(i_{A'} \otimes 1_B) \oplus (1_A \otimes i_{B'})} A \otimes B \xrightarrow{j_{A'} \otimes j_{B'}} A/A' \otimes B/B' \to 0$$

is exact, where $i_{A'}: A' \to A$ and $i_{B'}: B' \to B$ are the inclusion R-homomorphisms, and $j_{A'}: A \to A/A'$ and $j_{B'}: B \to B/B'$ are the induced canonical epimorphisms. It follows from the exactness of (1) that if

$$0 \longrightarrow A' \xrightarrow{f} A \xrightarrow{g} A'' \longrightarrow 0$$

and

$$0 \longrightarrow B' \xrightarrow{f'} B \xrightarrow{g'} B'' \longrightarrow 0$$

are exact sequences, then

(2) $$(A' \otimes B) \oplus (A \otimes B') \xrightarrow{(f \otimes 1_B) \oplus (1_A \otimes f')} A \otimes B \xrightarrow{g \otimes g'} A'' \otimes B'' \to 0$$

is an exact sequence. In particular, the exactness of the sequences

$$0 \to A' \xrightarrow{f} A \xrightarrow{g} A'' \to 0$$

and

$$0 \to 0 \to X \xrightarrow{1_X} X \to 0,$$

where X is an R-module, implies the exactness of the sequence

$$A' \otimes X \xrightarrow{f \otimes 1_X} A \otimes X \xrightarrow{g \otimes 1_X} A'' \otimes X \to 0.$$

7.3 *The homomorphism* $\mathrm{Hom}_R(f, g)$. When the ring R is not necessarily commutative the set $\mathrm{Hom}_R(A, B)$ of the R-homomorphisms from the R-module A to the R-module B is an abelian group. Let $f: A' \to A$, $g: B \to B'$ be R-homomorphisms. Define the map

$$\mathrm{Hom}_R(f, g): \mathrm{Hom}_R(A, B) \to \mathrm{Hom}_R(A', B')$$

by $\mathrm{Hom}_R(f,g)\varphi = g\varphi f$; $\mathrm{Hom}_R(f,g)$ is a homomorphism of abelian groups, for,

$$\begin{aligned}\mathrm{Hom}_R(f, g)(\varphi + \psi) &= g(\varphi + \psi)f = g\varphi f + g\psi f \\ &= \mathrm{Hom}_R(f, g)\varphi + \mathrm{Hom}_R(f, g)\psi.\end{aligned}$$

Moreover

$$\mathrm{Hom}_R(f_1 + f_2, g) = \mathrm{Hom}_R(f_1, g) + \mathrm{Hom}_R(f_2, g)$$
$$\mathrm{Hom}_R(f, g_1 + g_2) = \mathrm{Hom}_R(f, g_1) + \mathrm{Hom}_R(f, g_2).$$

If $A'' \xrightarrow{f'} A' \xrightarrow{f} A$ and $B \xrightarrow{g} B' \xrightarrow{g'} B''$ are R-homomorphisms, then

$$\mathrm{Hom}_R(f', g')\mathrm{Hom}_R(f, g)\varphi = \mathrm{Hom}_R(f', g')g\varphi f = g'g\varphi ff'.$$

Hence

$$\mathrm{Hom}_R(f', g')\mathrm{Hom}_R(f, g) = \mathrm{Hom}_R(ff', g'g).$$

7.4 *Let* $0 \to A \xrightarrow{i} B \xrightarrow{j} C \to 0$ *be an exact sequence of R-homomorphisms. If X is an R-module, then:*

(i) $$0 \to \text{Hom}_R(X, A) \xrightarrow{\text{Hom}_R(1_X, i)} \text{Hom}_R(X, B) \xrightarrow{\text{Hom}_R(1_X, j)} \text{Hom}_R(X, C)$$

is exact.

(ii) $$0 \to \text{Hom}_R(C, X) \xrightarrow{\text{Hom}_R(j, 1_X)} \text{Hom}_R(B, X) \xrightarrow{\text{Hom}_R(i, 1_X)} \text{Hom}_R(A, X)$$

is exact.

PROOF We will prove (ii). The proof of (i) is similar.

Exactness at $\text{Hom}_R(C, X)$. Let $f \in \text{Hom}_R(C, X)$ and suppose $\text{Hom}_R(j, 1_X)f = fj = 0$. Then $f = 0$ since j is an epimorphism.

Exactness at $\text{Hom}_R(B, X)$. Let $f \in \text{Hom}_R(C, X)$, then

$$\text{Hom}_R(i, 1_X)\text{Hom}_R(j, 1_X)f = fji = 0.$$

Hence $\text{im Hom}_R(j, 1_X) \subseteqq \ker \text{Hom}_R(i, 1_X)$.* To show $\text{im Hom}_R(j, 1_X) \supseteqq \ker \text{Hom}_R(i, 1_X)$, let $g \in \text{Hom}_R(B, X)$ and suppose $\text{Hom}_R(i, 1_X)g = gi = 0$. Let $f: C \to X$ be the map defined by $f(c) = g(b)$ where $b \in j^{-1}(c)$; f is well defined since if $b' \in j^{-1}(c)$, then $j(b - b') = 0$, and hence there exists $a \in A$ such that $ia = b - b'$. Thus $g(b - b') = gia = 0$. Therefore f is well defined. Furthermore, $f \in \text{Hom}_R(C, X)$ for $f(c_1 + c_2) = g(b_1 + b_2)$ where $b_1 \in j^{-1}(c_1)$ and $b_2 \in j^{-1}(c_2)$, and moreover

$$f(rc) = g(rb) = rg(b) = rf(c) \qquad \text{where} \quad b \in j^{-1}(c).$$

The identity $\text{Hom}_R(j, 1_X)f = fj = g$ establishes our assertion.

7.5 *Let* $0 \to A \xrightarrow{i} B \xrightarrow{j} C \to 0$ *be an exact sequence of R-homomorphisms. If iA is a direct summand of B* ($B = iA \oplus B'$ *for some R-module* B'), *then for any R-module X*

(i) *if R is commutative*

$$0 \to A \otimes_R X \xrightarrow{i \otimes 1_X} B \otimes_R X \xrightarrow{j \otimes 1_X} C \otimes_R X \to 0$$

is exact,

*im $\text{Hom}_R(j, 1_X)$ stands for the image of the homomorphism $\text{Hom}_R(j, 1_X)$; ker $\text{Hom}_R(i, 1_X)$ stands for the kernel of the homomorphism $\text{Hom}_R(i, 1_X)$.

(ii) $$0 \to \mathrm{Hom}_R(C, X) \xrightarrow{\mathrm{Hom}_R(j,1_X)} \mathrm{Hom}_R(B, X) \xrightarrow{\mathrm{Hom}_R(i,1_X)} \mathrm{Hom}_R(A, X) \to 0$$

is exact,

(iii) $$0 \to \mathrm{Hom}_R(X, A) \xrightarrow{\mathrm{Hom}_R(1_X,i)} \mathrm{Hom}_R(X, B) \xrightarrow{\mathrm{Hom}_R(1_X,j)} \mathrm{Hom}_R(X, C) \to 0$$

is exact.

Proof of (i) $B \otimes_R X = (iA \otimes X) \oplus B' \otimes X$ by 3.2. Hence $i \otimes 1_X$ is a monomorphism.

Proof of (ii) Let $f \in \mathrm{Hom}_R(A, X)$. Define $f' \in \mathrm{Hom}_R(iA, X)$ by $f'(ia) = f(a)$. Define $g \in \mathrm{Hom}_R(B, X)$ by

$$g = \begin{cases} f' & \text{on } iA \\ 0 & \text{on } B'. \end{cases}$$

Then $f = \mathrm{Hom}_R(i, 1_X)g$, for,

$$(\mathrm{Hom}_R(i, 1_X)g)(a) = (gi)(a) = f'(ia) = f(a).$$

Proof of (iii) Let $f \in \mathrm{Hom}_R(X, C)$. Suppose $ia + b' \in iA \oplus B'$ is such that $j(ia + b') = f(x)$. If $b \in j^{-1}(f(x))$, then $b - (ia + b') \in iA$. Thus if $ia_1 + b_1', ia_2 + b_2' \in j^{-1}(f(x))$ with $a_1, a_2 \in A$, $b_1', b_2' \in B'$, then $b_1' = b_2'$. Define $g: X \to B$ by $g(x) = b'$; g is an R-homomorphism, and moreover $\mathrm{Hom}_R(1_X, j)g = jg = f$.

7.6 The R-module P is called *projective* if given

$$\begin{array}{ccccc} & & P & & \\ & & \downarrow f & & \\ B & \xrightarrow{j} & C & \longrightarrow & 0 \end{array}$$

where the row is exact, there exists an R-homomorphism $g: P \to B$ such

that the diagram

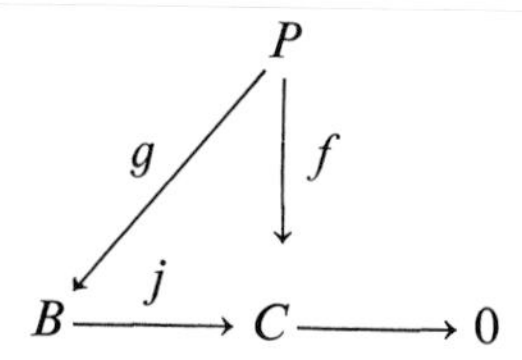

is commutative.

EXAMPLE Let $\mathbf{Z}$ be the integers and $\mathbf{Z}_n$ be the residue classes of integers modulo n. Then $\mathbf{Z}_2$ is not a projective $\mathbf{Z}$-module, for, given the diagram

$$\begin{array}{ccccc} & & \mathbf{Z}_2 & & \\ & & \downarrow{\scriptstyle 1_{\mathbf{Z}_2}} & & \\ \mathbf{Z}_4 & \xrightarrow{j} & \mathbf{Z}_2 & \longrightarrow & 0 \end{array}$$

where $j(1) = j(3) = 1$, $j(0) = j(2) = 0$, there is no $\mathbf{Z}$-homomorphism $g: \mathbf{Z}_2 \to \mathbf{Z}_4$ such that $jg = 1_{\mathbf{Z}_2}$. [If $g(1) = 1$ or $g(1) = 3$, then we would have $0 = 2$ in $\mathbf{Z}_4$; if $g(1) = 2$, then we would have $0 = 1$ in $\mathbf{Z}_2$.]

7.7 It follows from 7.6 that *if P is a projective R-module and $0 \to A \xrightarrow{i} B \xrightarrow{j} C \to 0$ is an exact sequence of R-homomorphisms, then*

$$0 \to \operatorname{Hom}_R(P, A) \xrightarrow{\operatorname{Hom}_R(1_X, i)} \operatorname{Hom}_R(P, B) \xrightarrow{\operatorname{Hom}_R(1_X, j)} \operatorname{Hom}_R(P, C) \to 0$$

is exact.

7.8 *Any free R-module is projective.*

PROOF Let $F = \sum_{\iota \in I} Rx_\iota$ be a free R-module generated by the symbols x_ι, $\iota \in I$. Given

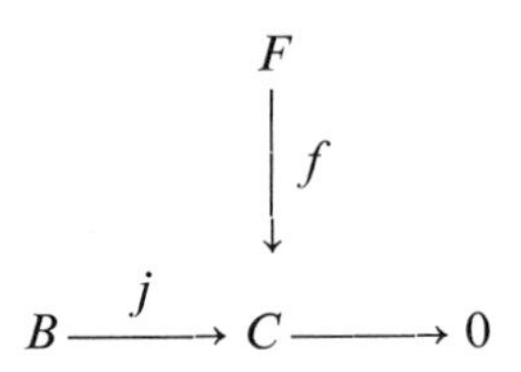

where the row is exact, let the set $\{b_\iota\}_{\iota \in I} \subseteq B$ be such that $j(b_\iota) = f(1x_\iota)$ for every $\iota \in I$. Define $g: F \to B$ by $g(1x_\iota) = b_\iota$, then $jg = f$ and g is an R-homomorphism (see 1.2, (ii)).

8. Homology and Cohomology

In this section the ring R is not necessarily commutative.

8.1 A *chain complex* of R-homomorphisms is a sequence

$$A_*: \cdots \longleftarrow A_{r-2} \xleftarrow{d_{r-1}} A_{r-1} \xleftarrow{d_r} A_r \longleftarrow \cdots$$

of R-homomorphisms such that

(1) $$d_{r-1}d_r = 0 \quad \text{for every integer} \quad r.$$

If the condition (1) is satisfied, we simply say A_* is a chain complex, and d_r is a *boundary* homomorphism.

A cochain complex is a sequence

$$A^*: \cdots \longrightarrow A^{r-1} \xrightarrow{d^{r-1}} A^r \xrightarrow{d^r} A^{r+1} \longrightarrow \cdots$$

of R-homomorphisms such that

(2) $$d^r d^{r-1} = 0 \quad \text{for every integer} \quad r.$$

d^r is called a coboundary homomorphism. The elements of the R-submodule $C_r = \{a \in A_r \text{ s.t.} d_r a = 0\}$* of A_r are called the *r-cycles* of the chain complex A_*. The elements of the R-submodule

$$B_r = \{a \in A_r \text{ s.t.} a = d_{r+1}a' \text{ for some } a' \in A_{r+1}\}$$

*s.t. stands for *such that*.

of A_r are called *r-boundaries* of the chain complex A_*. The chain complex A_* is *acyclic* (or exact) if and only if $B_r = C_r$ for every subscript r. In general, however, B_r is a submodule of C_r. The R-module $C_r/B_r = H_r(A_*)$ is called the rth *homology* of the chain complex A_*. The R-module $H_r(A_*)$ measures the amount by which exactness is violated at A_r in the complex A_*. Similarly $C^r = \{a \in A^r \text{ s.t. } d^r a = 0\}$ is the R-module of *r-cocycles* of A^*, and

$$B^r = \{a \in A^r \text{ s.t. } a = d^{r-1}a' \text{ for some } a' \in A^{r-1}\}$$

is the R-module of *r-coboundaries* of A^*. The R-module $C^r/B^r = H^r(A^*)$ is the rth *cohomology* of the cochain complex A^*. From a given chain complex A_* of R-homomorphisms we can construct other complexes. Let X be an R-module. View A_r and X as Z-modules (Z is the integers) and d_r as a Z-homomorphism. Then $A_r \otimes_Z X$ makes sense and we can construct the chain complex of Z-homomorphisms

$$A_* \otimes_Z X\colon \cdots \leftarrow A_{r-1} \otimes_Z X \xleftarrow{d_r \otimes 1_X} A_r \otimes_Z X \leftarrow \cdots$$

Observe that

$$(d_r \otimes 1_X)(d_{r+1} \otimes 1_X) = d_r d_{r+1} \otimes 1_X = 0$$

by 4.2. We may also construct the cochain complex of Z-homomorphisms

$$\mathrm{Hom}_R(A_*, X)\colon \cdots \to \mathrm{Hom}_R(A_{r-1}, X) \xrightarrow{\mathrm{Hom}_R(d_r, 1_X)} \mathrm{Hom}_R(A_r, X) \to \cdots.$$

Observe that

$$\mathrm{Hom}_R(d_{r+1}, 1_X)\mathrm{Hom}_R(d_r, 1_X) = \mathrm{Hom}_R(d_r d_{r+1}, 1_X) = 0$$

by 7.3.

8.2. Let A_*, A'_* be chain complexes of R-homomorphisms. By a *morphism* $f\colon A_* \to A'_*$ *of complexes* we shall mean a commutative diagram

$$\begin{array}{ccccccccc}
\cdots & \longleftarrow & A_{r-1} & \xleftarrow{d_r} & A_r & \xleftarrow{d_{r+1}} & A_{r+1} & \longleftarrow & \cdots \\
 & & \downarrow f_{r-1} & & \downarrow f_r & & \downarrow f_{r+1} & & \\
\cdots & \longleftarrow & A'_{r-1} & \xleftarrow{d'_r} & A'_r & \xleftarrow{d'_{r+1}} & A'_{r+1} & \longleftarrow & \cdots
\end{array}$$

where $f_r: A_r \to A'_r$ is an R-homomorphism for every r. We may informally refer to the commutativity of this diagram by writing $fd = d'f$. Each morphism $f: A_* \to A'_*$ of complexes induces an R-homomorphism $\bar{f}_r: H_r(A_*) \to H_r(A'_*)$ for every r, defined by $\bar{f}_r(c_r + B_r) = f_r c_r + B_r$ where c_r is an r-cycle, $B_r = d_{r+1}A_{r+1}$ is the R-module of r-boundaries of A_*, and $B'_r = d'_{r+1}A'_{r+1}$. The symbol $\bar{f}_r$ is indeed a mapping into $H_r(A'_*)$ since

$$d'_r f_r c_r = f_{r-1} d_r c_r = f_{r-1}(0) = 0,$$

i.e., $f_r(c_r)$ is an r-cycle. To show $\bar{f}_r$ is well defined represent the element $c_r + B_r$ of $H_r(A_*)$ by $c_r + d_{r+1}a_{r+1} + B_r$. Then

$$\begin{aligned}\bar{f}_r(c_r + d_{r+1}a_{r+1} + B_r) &= f_r c_r + f_r d_{r+1}a_{r+1} + B'_r \\ &= f_r c_r + d'_{r+1} f_{r+1} a_{r+1} + B'_r \\ &= f_r c_r + B'_r.\end{aligned}$$

Hence $\bar{f}_r$ is well defined. Since f_r is an R-homomorphism, it follows that $\bar{f}_r$ is an R-homomorphism.

8.3 *The connecting homomorphism.* We say the sequence

$$0 \to A'_* \xrightarrow{i} A_* \xrightarrow{j} A''_* \to 0$$

of morphisms of chain complexes is exact if the sequence

$$0 \longrightarrow A'_r \xrightarrow{i_r} A_r \xrightarrow{j_r} A''_r \longrightarrow 0$$

of R-homomorphisms is exact for every integer r. Define the R-homomorphism $\bar{d}_r: H_r(A''_*) \to H_{r-1}(A'_*)$ called the *connecting homomorphism by*

$$\bar{d}_r(a''_r + B''_r) = a'_{r-1} + B'_{r-1}$$

where B''_r, B'_{r-1}, are the rth and $(r-1)$st boundaries of the complexes A''_* and A'_*, respectively, and where a'_{r-1} is obtained as follows:

(1) Let $a_r \in j_r^{-1}(a''_r)$.

(2) Since $j_{r-1}(d_r a_r) = d''_r j_r a_r = d''_r a''_r = 0$, there exists $a'_{r-1} \in A'_{r-1}$ such that $i_{r-1}a'_{r-1} = d_r a_r$.

(3) a'_{r-1} is a cycle for

$$i_{r-2} d'_{r-1} a'_{r-1} = d_{r-1} i_{r-1} a'_{r-1} = d_{r-1} d_r a_r = 0$$

and i_{r-2} is a monomorphism.

(4) $a'_{r-1} + B'_{r-1}$ is independent of the representation $a''_r + B''_r$ and the choice of a_r. For, let $a''_r + d''_{r+1}a''_{r+1} + B''_r$ be another representation of $a''_r + B''_r$. Then in (1), (2), and (3) replace a''_r by $a''_r + d''_{r+1}a''_{r+1}$, and if $a''_{r+1} = j_{r+1}a_{r+1}$, replace a_r by $a_r + d_{r+1}a_{r+1} + ia'_r$. Then a'_{r-1} is replaced by $a'_{r-1} + d'_r a'_r$.

The verification that $\bar{d}_r$ is an R-homomorphism is done by the constructions in (1) and (2).

8.4 *If the sequence* $0 \to A'_* \xrightarrow{i} A_* \xrightarrow{j} A''_* \to 0$ *of morphisms of complexes is exact, then the long sequence*

$$\cdots \xrightarrow{\bar{j}_{r+1}} H_{r+1}(A''_*) \xrightarrow{\bar{d}_{r+1}} H_r(A'_*) \xrightarrow{\bar{i}_r} H_r(A_*) \xrightarrow{\bar{j}_r} H_r(A''_*) \longrightarrow \cdots$$

of R-homomorphisms is exact.

PROOF Computations (i)–(iii) below establish that the long sequence is a complex, and computations (iv)–(vi) complete the proof.

(i)
$$\begin{aligned} \bar{j}_r\bar{i}_r(a'_r + B'_r) &= \bar{j}_r(i_r a'_r + B_r) \\ &= j_r i_r a'_r + B''_r \\ &= B''_r \end{aligned}$$

(ii)
$$\bar{d}_r\bar{j}_r(a_r + B_r) = \bar{d}_r(j_r a_r + B''_r) = a'_{r-1} + B'_r$$

where $i_r a'_{r-1} = d_r a_r$. But since a_r is a cycle, $a'_{r-1} = 0$. Thus $\bar{d}_r\bar{j}_r(a_r + B_r) = B'_r$.

(iii)
$$\bar{i}_r\bar{d}_{r+1}(a''_{r+1} + B''_{r+1}) = \bar{i}_r(a'_r + B'_r)$$

where if $a_{r+1} \in j_{r+1}^{-1}(a''_{r+1})$, then $i_r a'_r = d_{r+1}a_{r+1}$. Therefore

$$\bar{i}_r\bar{d}_{r+1}(a''_{r+1} + B''_{r+1}) = d_{r+1}a_{r+1} + B_r = B_r.$$

(iv) Suppose $\bar{i}_r(a'_r + B'_r) = B_r$, then $i_r a'_r = d_{r+1}a_{r+1}$ for some $a_{r+1} \in A_{r+1}$. Moreover $j_{r+1}a_{r+1}$ is a cycle since

$$d''_{r+1}j_{r+1}a_{r+1} = j_r d_{r+1}a_{r+1} = j_r i_r a'_r = 0.$$

It follows from the definition of $\bar{d}_r$ that

$$\bar{d}_r(j_{r+1}a_{r+1} + B''_{r+1}) = a'_r + B'_r.$$

(v) Suppose $\bar{j}_r(a_r + B_r) = B''_r$, then $j_r a_r = d''_{r+1} a''_{r+1}$ for some $a''_{r+1} \in A''_{r+1}$. If a_{r+1} is such that $j_{r+1} a_{r+1} = a''_{r+1}$, then $j_r(a_r - d_{r+1} a_{r+1}) = 0$. Thus there exists $a'_r \in A'_r$ such that $i_r a'_r = a_r - d_{r+1} a_{r+1}$; a'_r is a cycle since $i_{r-1} d'_r a'_r = d_r i_r a'_r = 0$. Hence

$$\bar{\imath}_r(a'_r + B'_r) = a_r - d_{r+1} a_{r+1} + B_r = a_r + B_r.$$

(vi) Suppose $\bar{d}_r(a''_r + B''_r) = B'_{r-1}$. Then if $a''_r = j_r a_r$, we have $d_r a_r = i_{r-1} d'_r a'_r$ for some $a'_r \in A'_r$.

It follows that $a_r - i_r a'_r$ is a cycle, moreover $j_r(a_r - i_r a'_r) = a''_r$. Thus

$$\bar{j}_r(a_r - i_r a'_r + B_r) = a''_r + B''_r.$$

8.5 *Let $f\colon X_* \to X'_*$, $g\colon Y_* \to Y'_*$, $h\colon Z_* \to Z'_*$ be morphisms of chain complexes. If the commutative diagram*

$$\begin{array}{ccccccccc} 0 & \longrightarrow & X_* & \xrightarrow{i} & Y_* & \xrightarrow{j} & Z_* & \longrightarrow & 0 \\ & & \downarrow f & & \downarrow g & & \downarrow h & & \\ 0 & \longrightarrow & X'_* & \xrightarrow{i'} & Y'_* & \xrightarrow{j'} & Z'_* & \longrightarrow & 0 \end{array}$$

has exact rows, then the diagram

$$\begin{array}{ccccccccc} \cdots \longrightarrow & H_{r+1}(Z_*) & \xrightarrow{\bar{d}_{r+1}} & H_r(X_*) & \xrightarrow{\bar{\imath}_r} & H_r(Y_*) & \xrightarrow{\bar{j}_r} & H_r(Z_*) & \longrightarrow \cdots \\ & \downarrow \bar{h}_{r+1} & & \downarrow \bar{f}_r & & \downarrow \bar{g}_r & & \downarrow \bar{h}_r & \\ \cdots \longrightarrow & H_{r+1}(Z'_*) & \xrightarrow{\bar{d}'_{r+1}} & H_r(X'_*) & \xrightarrow{\bar{\imath}'_r} & H_r(Y'_*) & \xrightarrow{\bar{j}'_r} & H_r(Z'_*) & \longrightarrow \cdots \end{array}$$

is commutative.

Proof $\bar{g}_r \bar{\imath}_r = \bar{\imath}'_r \bar{f}_r$ and $\bar{j}'_r \bar{g}_r = \bar{h}_r \bar{j}_r$ for every r follow from the identities $g_r i_r = i'_r f_r$ and $j'_r g_r = h_r j_r$. To show $\bar{d}'_r \bar{h}_r = \bar{f}_{r-1} \bar{d}_r$, let

$$\bar{d}_r(z_r + d_{r+1} Z_{r+1}) = x_{r-1} + d_r X_r.$$

Then $h_r z_r = h_r j_r y_r = j'_r g_r y_r$ for some $y_r \in Y_r$. Therefore

(1) $$\bar{d}'_r\bar{h}_r(z_r + d_{r+1}Z_{r+1}) = \bar{d}'_r(h_r z_r + d'_{r+1}Z'_{r+1})$$
$$= f_{r-1}x_{r-1} + d'_r X'_r$$

because

$$d'_r g_r y_r = g_{r-1}d_r y_r = g_{r-1}i_{r-1}x_{r-1} = i'_{r-1}f_{r-1}x_{r-1},$$

and

(2) $$\bar{f}_{r-1}\bar{d}_r(z_r + d_{r+1}Z_{r+1}) = \bar{f}_{r-1}(x_{r-1} + d_r X_r)$$
$$= f_{r-1}x_{r-1} + d'_r X'_r.$$

Comparing (1) and (2) we have $\bar{d}'_r\bar{h}_r = \bar{f}_{r-1}\bar{d}_r$.

9. Homotopy

9.1 Let $f,\ g: X_* \rightrightarrows X'_*$ be morphisms of complexes of R-homomorphisms. Then $f + g: X_* \to X'_*$ defined by $(f+g)_r = f_r + g_r$ is a morphism of complexes. Moreover $(\overline{f+g})_r = \bar{f}_r + \bar{g}_r$. We say the morphisms of complexes f and g are *homotopic* if there exists a family $\{D_r: X_r \to X'_{r+1}\}_{r\in Z}$ of R-homomorphisms called *splitting homotopies* such that

$$f_r - g_r = d'_{r+1}D_r + D_{r-1}d_r.$$

The R-homomorphisms involved are displayed diagrammatically as follows:

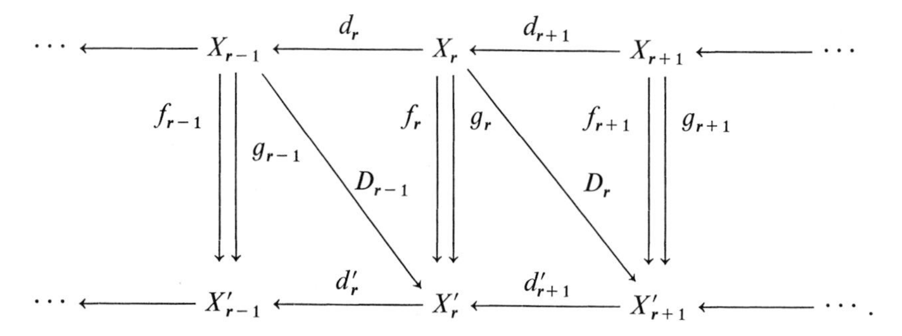

9.2 *If $f,\ g: X_* \rightrightarrows X'_*$ are homotopic morphisms of complexes, then $\bar{f} = \bar{g}$.*

Proof Let $x_r + d_{r+1}X_{r+1} \in H_r(X_*)$. Then

$$\begin{aligned}(\overline{f - g})_r(x_r + d_{r+1}X_{r+1}) &= (f_r - g_r)x_r + d'_{r+1}X'_{r+1} \\ &= d'_{r+1}D_r x_r + D_{r-1}d_r x_r + d'_{r+1}X'_{r+1} \\ &= d'_{r+1}D_r x_r + d'_{r+1}X'_{r+1} \qquad \text{since} \quad x_r \text{ is a cycle} \\ &= d'_{r+1}X'_{r+1} \quad \text{since} \quad D_r x_r \in X'_{r+1}.\end{aligned}$$

Hence $\bar{f} = \bar{g}$ by 9.1.

9.3 *If the identity morphism $1_{X_*}: X_* \to X_*$ is homotopic to the trivial morphism, i.e., $1_{X_r} = d_{r+1}D_r + D_{r-1}d_r$, for every r, then $H_r(X_*) = 0$ for every r.*

Let x_r be an r-cycle, then

$$x_r = d_{r+1}D_r x_r + D_{r-1}d_r x_r = d_{r+1}D_r x_r$$

is an r-boundary.

In the special case where $R = \mathbf{Z}$ and X_r are free $\mathbf{Z}$-modules the converse of 9.3 holds. In fact:

9.4 *If X_* is an acyclic complex of homomorphisms of free abelian groups, there exist homomorphisms*

$$D_r: X_r \to X_{r+1} \qquad \textit{for every} \quad r$$

such that $1_{X_r} = d_{r+1}D_r + D_{r-1}d_r$.

Proof $d_r X_r$ is a free abelian group since it is a subgroup of the free abelian group X_{r-1}. Let $\{d_r x_r^{(\iota)}\}_{\iota \in I}$ be a basis of $d_r X_r$. Define the homomorphism $S_{r-1}: d_r X_r \to X_r$ by $S_{r-1}(d_r x_r^{(\iota)}) = x_r^{(\iota)}$ for every $\iota \in I$. Let $d_r x_r \in d_r X_r$, then $d_r x_r = \sum z_\iota d_r x_r^{(\iota)}$ where z_ι are integers, and

$$d_r S_{r-1}(d_r x_r) = \sum z_\iota d_r S_{r-1} d_r x_r^{(\iota)} = \sum z_\iota\, d_r x_r^{(\iota)} = d_r x_r.$$

Thus for any $x_r \in X_r$, $x_r - S_{r-1}d_r x_r \in d_{r+1}X_{r+1}$ since $d_r(x_r - S_{r-1}d_r x_r) = 0$ and $H_r(X_*) = 0$. Thus $S_r(x_r - S_{r-1}d_r x_r) \in X_{r+1}$. Define $D_r: X_r \to X_{r+1}$ by $D_r x_r = S_r(x_r - S_{r-1}d_r x_r)$. Then

(i) $$d_{r+1}D_r x_r = x_r - S_{r-1}d_r x_r$$

and

$$\text{(ii)} \qquad D_{r-1}d_rx_r = S_{r-1}(d_rx_r - S_{r-2}\,d_{r-1}d_rx_r) = S_{r-1}d_rx_r.$$

By (i) and (ii),

$$(d_{r+1}D_r + D_{r-1}d_r)(x_r) = x_r.$$

10. Direct and Inverse Systems

A partially ordered set $\{D, \leqq\}$ which satisfies the condition:

$\mathscr{D}$ *For any* $i, j \in D$ *there exists* $k \in D$ *such that* $i \leqq k$, $j \leqq k$,
is called a *directed* set.

A directed set D may be viewed as a category whose objects are the elements of the set D and whose morphisms are

$$\operatorname{Hom}(i, j) = \begin{cases} \varnothing & \text{if } i \nleqq j \\ \text{the singleton "} i \leqq j \text{"} & \text{if } i \leqq j, \end{cases}$$

such that the condition $\mathscr{D}$ is satisfied.

Let $\mathscr{C}$ be a category. A direct system in the category $\mathscr{C}$ is a functor from a directed set D to $\mathscr{C}$. An inverse system is a contravariant functor from D to $\mathscr{C}$. We say the *constant functor* C [i.e., $C(i)$ is the object C of $\mathscr{C}$ for every $i \in D$, and for any morphism $i \leqq j$ in D, $C(i \leqq j) = 1_C$] is a *limit* of the direct system $T: D \to \mathscr{C}$ if there is a natural transformation $\alpha: T \to C$ (i.e., for any $i \in D$, there is given a morphism $\alpha_i: T_i \to C$ such that the diagram

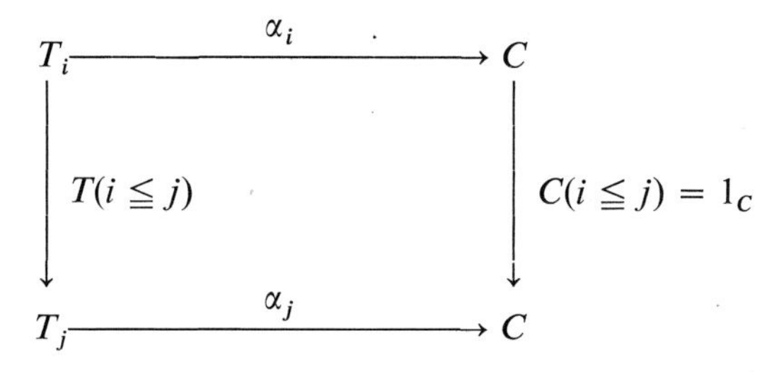

is commutative) such that given any constant functor $B: D \to \mathscr{C}$ and a natural transformation $\beta: T \to B$, there is a unique natural transformation $\bar{\beta}: C \to B$ such that $\bar{\beta} \circ \alpha = \beta$. We observe that the natural transformation $\bar{\beta}$ must necessarily satisfy $\bar{\beta}_i = \bar{\beta}_j$ for every $i, j \in D$. For, there exists k with $i \leqq k, j \leqq k$ such that

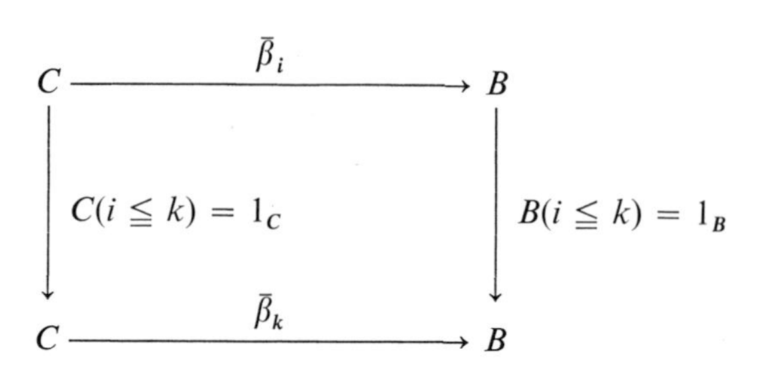

is commutative, and hence $\bar{\beta}_i = \bar{\beta}_k$. Similarly $\bar{\beta}_j = \bar{\beta}_k$. In other words, the natural transformation α satisfies the universal mapping property with respect to natural transformations from the functor T to constant functors. If C and B above are both limits of the direct system $T: D \to \mathscr{C}$, then the unique natural transformation $\bar{\beta}: C \to B$ above is an isomorphism of constant functors, i.e., it is an isomorphism in the category $\mathscr{C}$. Hence the limit C of the direct system T is determined up to a canonical isomorphism.

Direct systems and limits can be defined in terms of the internal properties of the category $\mathscr{C}$. For our purposes we assume $\mathscr{C}$ is the category of abelian groups. The family of homomorphisms

$$\{\rho^i_j: A_i \to A_j, \quad i, j \in D, \quad i \leqq j\}$$

is a direct system if D is a directed set and

$$\rho^i_i = 1_{A_i} \qquad \text{for every} \quad i$$

$$\rho^i_j \rho^j_k = \rho^i_k \qquad \text{for} \quad i \leqq j \leqq k.$$

Consider the direct sum $\sum_{i \in D} A_i$. Let $\iota_i: A_i \to \sum_{i \in D} A_i$ be the inclusion of the ith summand. Introduce a relation in this direct sum by setting $\iota_j \rho^i_j a_i$ equivalent to $\iota_i a_i$ for $i \leqq j$ and $a_i \in A_i$. The factor group of $\sum_{i \in D} A_i$ modulo this equivalence relation is $A = \sum_{i \in D} A_i / R$ where R is the subgroup of $\sum_{i \in D} A_i$ generated by all the elements of the form

$$\iota_j \rho^i_j a_i - \iota_i a_i \qquad \text{for} \quad i \leqq j.$$

A is the *limit* of the direct system

$$\{\rho_j^i: A_i \to A_j, i, j \in D, i \leqq j\}.$$

This assertion follows from the results we obtain below.

The canonical homomorphism $\sum_{i \in D} A_i \to A$ defines for each $i \in D$ the homomorphism

$$\rho_i = \rho \iota_i.$$

Since ρ maps each relation $\iota_j \rho_j^i a_i - \iota_i a_i$ to zero, we have

(1) $$\rho_j \rho_j^i = \rho_i.$$

10.1 *For any $a \in A$ there is an $i \in D$ and $a_i \in A_i$ such that $\rho_i a_i = a$.*

PROOF There exists $p \in \sum_{i \in D} A_i$ such that $\rho p = a$, $p = \iota_{i_1} p_{i_1} + \cdots + \iota_{i_n} p_{i_n}$ where $p_{i_\kappa} \in A_{i_\kappa}$. Since D is directed, there exists $i \in D$ such that $i_\kappa \leqq i$ for $\kappa = 1, \ldots, n$. Let $q_\kappa = \rho_i^{i_\kappa} p_{i_\kappa} \in A_i$ and $a_i = q_1 + \cdots + q_n$. Since $\iota_i q_\kappa - \iota_{i_\kappa} p_{i_\kappa} \in R$, it follows that $\iota_i a_i - p \in R$. Hence $\rho_i a_i - a = \rho(\iota_i a_i - p) = 0$, so $\rho_i a_i = a$.

10.2 *If $a \in A_k$ and $\rho_k a = 0$, there exists $l \in D$, $k \leq l$, such that $\rho_l^k a = 0$.*

PROOF $\iota_k a$ is a linear combination of elements of the form

(2) $$\iota_j \rho_j^i a_i - \iota_i a_i$$

with integer coefficients. Since any integer multiple of an element of the form (2) is again of the form (2), $\iota_k a$ is a sum of elements of the form (2). Let $l \in D$ be such that $k \leqq l$ and $i, j \leqq l$ for every i, j present in the expression for $\iota_k a$. Then

$$\iota_l \rho_l^k a = (\iota_l \rho_l^k a - \iota_k a) + \iota_k a$$

is also a sum of elements of type (2) and l exceeds k and all i, j implicitly present in the right-hand side. If $\iota_j \rho_j^i a_i - \iota_i a_i$ is a summand in $\iota_k a$, we can write

$$\iota_j \rho_j^i a_i - \iota_i a_i = (\iota_l \rho_l^i a_i - \iota_i a_i) + (\iota_l \rho_l^j (-\rho_j^i a_i) - \iota_j (-\rho_j^i a_j)).$$

Thus

$$\imath_l \rho_l^k a = \sum_i (\imath_l \rho_l^i a_i - \imath_i a_i),$$

i.e.,

$$\imath_l \rho_l^k a - \sum_i (\imath_l \rho_l^i a_i - \imath_i a_i) = 0.$$

Since the left-hand side is an element of the direct sum $\sum_{j \in D} A_j$, and is zero, the individual summands are zero. Hence $a_i = 0$ for all $i \neq l$ and for $i = l$, $\imath_l \rho_l^l a_l - \imath_l a_l = 0$. Thus $\imath_l \rho_l^k a = 0$, and since $\imath_l$ is a monomorphism, this implies $\rho_l^k a = 0$.

10.3 *$\rho_i a_i = \rho_j a_j$ if and only if there exists k, $i, j \leqq k$, such that $\rho_k^i a_i = \rho_k^j a_j$.*

Proof To show the sufficiency observe that $\imath_k \rho_k^i a_i - \imath_i a_i$ and $\imath_k \rho_k^j a_j - \imath_j a_j$ are elements in R. Hence their difference $\imath_i a_i - \imath_j a_j$ is an element of R. Therefore

$$\rho(\imath_i a_i - \imath_j a_j) = \rho_i a_i - \rho_j a_j = 0.$$

To show the necessity, choose l such that $i, j \leqq l$ and let

$$a = \rho_l^i a_i - \rho_l^j a_j. \tag{3}$$

Then

$$\rho_l a = \rho_l \rho_l^i a_i - \rho_l \rho_l^j a_j = \rho_i a_i - \rho_j a_j = 0.$$

Hence, by 10.2, there exists $k \in D$, $l \leqq k$, such that $\rho_k^l a = 0$. Therefore the application of ρ_k^l to the identity (3) results in the identity

$$\rho_k^i a_i - \rho_k^j a_j = 0.$$

10.4 *If $\{\rho_j^i: A_i \to A_j;\ i, j \in D,\ i \leqq j\}$ is a direct system of abelian groups with limit A, and the homomorphisms ρ_j^i are monomorphisms, then the homomorphisms $\rho_i: A_i \to A$ are monomorphisms.*

Proof Follows trivially from 10.2.

Let

$$\{\rho_j^i\colon A_i \to A_j;\ i, j \in D,\ i \leqq j\}$$

and

$$\{\rho_j'^i\colon A_i' \to A_j';\ i, j \in D,\ i \leqq j\}$$

be direct systems with limits A and A', respectively. Let $\{\varphi_i\colon A_i \to A_i';\ i \in D\}$ be a family of maps such that for $i \leqq j$, the diagram

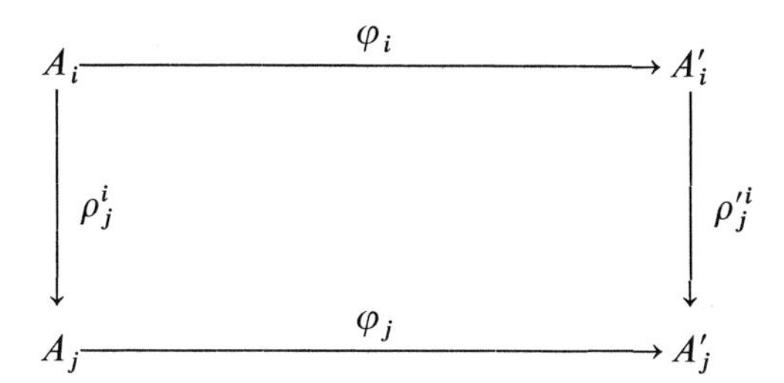

is commutative. By the universal mapping property of limits the family of maps $\{\rho_i'\varphi_i\colon A_i \to A';\ i \in D\}$ induces a unique map $\varphi\colon A \to A'$ such that $\rho_i'\varphi_i = \varphi\rho_i$.

10.5 *If the maps φ_i are monomorphisms, then φ is a monomorphism.*

Proof Let $a \in A$ such that $\varphi a = 0$. By 10.1, there is an $i \in D$ and $a_i \in A_i$ such that $\rho_i a_i = a$. Then $\rho_i'\varphi_i a_i = 0$ and, by 10.2, there exists $j \in D$, $i \leqq j$, such that $\rho'^i_j\varphi_i a_i = 0$. Hence $\varphi_j\rho_j^i a_i = 0$, and since φ_j is a monomorphism, this implies $a = \rho_j\rho_j^i a_i = 0$.

10.6 *If the maps φ_i in 10.5 are epimorphisms, then φ is an epimorphism.*

Proof Let $a' \in A'$. By 10.1, there is an $i \in D$ and $a_i' \in A_i'$ such that $\rho_i' a_i' = a'$. Since φ_i is an epimorphism, there exists an $a_i \in A_i$ such that $\varphi_i a_i = a_i'$, and by the definition of φ, $\varphi\rho_i a_i = \rho_i'\varphi_i a_i = a'$. Hence φ is an epimorphism.

We conclude this section with a brief discussion of inverse limits. The proofs of the statements below are similar to those of 10.1–10.6 and are left to the reader.

Let

$$\{\rho_i^j: A_j \to A_i; i, j \in D, i \leqq j\}$$

be an inverse system of abelian groups. The inverse limit A of this system is the subgroup of the product $\prod_{i \in D} A_i$ such that $(x_i)_{i \in D} \in A$ if for $i \leq j$, $\rho_i^j x_j = x_i$. Define the projection maps $\rho_k: A \to A_k$ by $\rho_k((x_i)_{i \in D}) = x_k$. Then for $i \leqq j$,

$$\rho_i = \rho_i^j \rho_j.$$

Let

$$\{\rho_i^j: A_j \to A_i; i, j \in D, i \leqq j\}$$

and

$$\{\rho_i'^j: A_j' \to A_i'; i, j \in D, i \leqq j\}$$

be inverse systems with limits A and A', respectively. Let

$$\{\varphi_i: A_i \to A_i', i \in D\}$$

be a family of maps such that for $i \leqq j$, the diagram

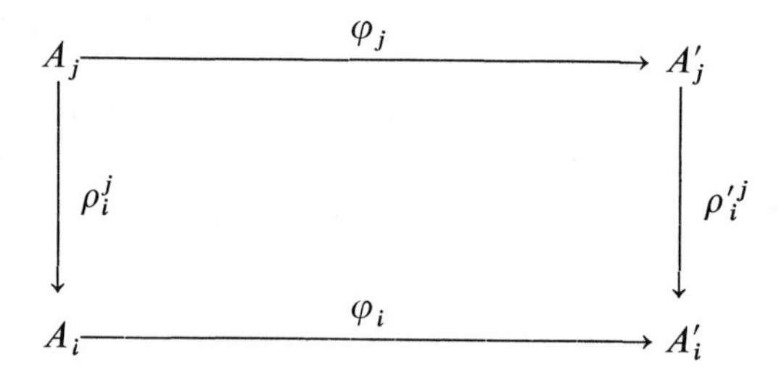

is commutative. The family of maps $\{\varphi_i \rho_i\}_{i \in D}$ induces a unique map $\varphi: A \to A'$. Moreover, if φ_i is an isomorphism for every $i \in D$, then φ is an isomorphism.

EXERCISES

1. (The Five Lemma) Given the row-exact commutative diagram of R-homomorphisms

$$\begin{array}{ccccccccc} A_1 & \longrightarrow & A_2 & \longrightarrow & A_3 & \longrightarrow & A_4 & \longrightarrow & A_5 \\ \downarrow f_1 & & \downarrow f_2 & & \downarrow f_3 & & \downarrow f_4 & & \downarrow f_5 \\ B_1 & \longrightarrow & B_2 & \longrightarrow & B_3 & \longrightarrow & B_4 & \longrightarrow & B_5 \end{array}$$

verify the following statements:

(i) If f_1 is an epimorphism, and f_2, f_4 are monomorphisms, then f_3 is a monomorphism.

(ii) If f_2, f_4 are epimorphisms, and f_5 is a monomorphism, then f_3 is an epimorphism.

2. (The Nine Lemma) Consider the commutative diagram

$$\begin{array}{ccccccccc} & & 0 & & 0 & & 0 & & \\ & & \downarrow & & \downarrow & & \downarrow & & \\ 0 & \to & A_1 & \to & A_2 & \to & A_3 & \to & 0 \\ & & \downarrow & & \downarrow & & \downarrow & & \\ 0 & \to & B_1 & \to & B_2 & \to & B_3 & \to & 0 \\ & & \downarrow & & \downarrow & & \downarrow & & \\ 0 & \to & C_1 & \to & C_2 & \to & C_3 & \to & 0 \\ & & \downarrow & & \downarrow & & \downarrow & & \\ & & 0 & & 0 & & 0 & & \end{array}$$

where the second row and all three columns are exact. Verify that the first row is exact if and only if the third row is exact.

CHAPTER

2

Cohomology of Finite Groups

In this chapter G is a multiplicative finite group, and $\mathbf{Z}$ is the ring of integers.

11. Group Rings

11.1 Let $\mathbf{Z}G$ be the set of all formal sums $\sum_{g\in G} z_g g$ where $z_g \in \mathbf{Z}$. We will consider two such formal sums $\sum_{g\in G} z_g g$ and $\sum_{h\in G} z'_h h$ equal if and only if $g = h$ implies $z_g = z'_h$. Define an addition in $\mathbf{Z}G$ by

$$\sum_{g\in G} z_g g + \sum_{g\in G} z'_g g = \sum_{g\in G} (z_g + z'_g)g$$

and a multiplication by

$$(\sum_{g\in G} z_g g)(\sum_{g'\in G} z'_{g'} g') = \sum_{g''\in G} z''_{g''} g''$$

where

$$z''_{g''} = \sum_{g\in G} z_{g'} \cdot z'_{g^{-1}\cdot g''}.$$

G together with these compositions is a ring called the *group ring* of G. Thus the elements $1g$, $g \in G$, are linearly independent and make up a basis of the free module structure of $\mathbf{Z}G$ over $\mathbf{Z}$. The mapping $\theta\colon G \to \mathbf{Z}G$ defined by $\theta g = 1g$ is a monomorphism of G into $\mathbf{Z}G$ [θ is one to one, and $\theta(gg') = \theta(g)\theta(g')$]. Moreover θ satisfies the following universal mapping property: If $\varphi\colon G \to R$ is any mapping of the group G into a ring R with identity, such that $\varphi(1) = 1$ and $\varphi(gg') = \varphi(g)\varphi(g')$, then there exists a unique homomorphism of rings $\bar{\varphi}\colon \mathbf{Z}G \to R$ such that the diagram

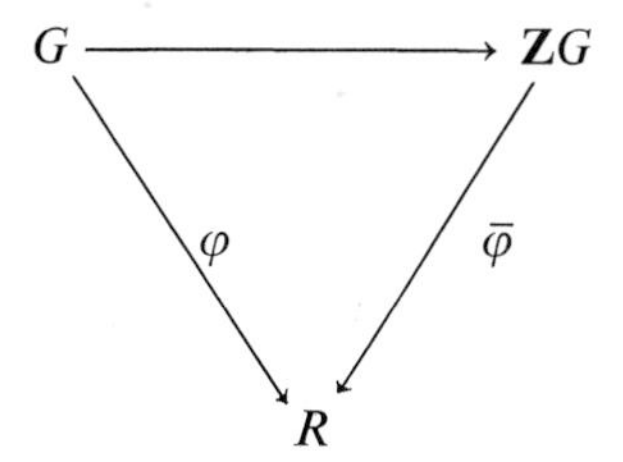

is commutative; the homomorphism $\bar{\varphi}$ is defined by

$$\bar{\varphi} \sum_{g \in G} z_g g = \sum_{g \in G} z_g \varphi(g).$$

11.2 Let A be an abelian group; we say A is a *G-module* if G operates on A, i.e., a composition $\kappa\colon G \times A \to A$ is given where $\kappa(g, a)$ is written ga such that:

(i) $1a = a$

(ii) $g(a_1 + a_2) = ga_1 + ga_2$

(iii) $(gg')a = g(g'a)$.

We will say κ is an *action* of G on A.

If A is a G-module, define the action of $\mathbf{Z}G$ on A by

$$\left(\sum_{g \in G} z_g g\right)a = \sum_{g \in G} z_g ga.$$

This action furnishes A with the structure of a $\mathbf{Z}G$-module.

EXAMPLE Let K be a finite field. Let G be the group of automorphisms of K. Then G operates on K^+ (the additive group of K).

12. The Trace Map

$S \leqq G$ will mean S is a subgroup of the group G. $S \trianglelefteq G$ will mean S is a normal subgroup of the group G.

12.1 Let $S \leqq G$. A G-module M is also an S-module where the action of S (or $\mathbf{Z}S$) on M is given by

$$S \times M \xrightarrow{i \times 1} G \times M \xrightarrow{\kappa} M$$

where $i: S \to G$ is the inclusion of S in G and κ is the action of G on M. Define

$$M^S = \{m: m \in M, sm = m \text{ for every } s \in S\}.$$

M^S is an abelian group. In general M^S is not a G-module; however, if $S \trianglelefteq G$, there exists an action $\kappa': G \times M^S \to M^S$ of G on M^S defined by $\kappa'(g, m) = \kappa(g, m) = gm$. To show $\kappa'(g, m) \in M^S$ for every $m \in M^S$, let $s \in S$. Then

$$s(\kappa'(g, m)) = s(gm) = (sg)m = (gs')m = gm = \kappa'(g, m).$$

Define the map $\bar{\kappa}: G/S \times M^S \to M^S$ by $\bar{\kappa}(gS, m) = gm$. It is easy to verify that conditions (i)–(iii) of 11.2 are satisfied. Therefore $S \trianglelefteq G$ implies M^S is a G/S-module.

Observe that we are imitating Galois theory in the sense that in Galois theory we are given a finite normal separable extension K of a field k. Let G be the (finite) group of the k-automorphisms of K, and let $K^\times$ be the multiplicative group of K. Then G operates on $K^\times$. Moreover if $S \trianglelefteq G$ and $(K^\times)^S = (K^S)^\times$ is the multiplicative group of the intermediate field K^S corresponding to S, then G/S is the k-automorphism group of K^S. Thus $(K^\times)^S$ is a G/S-module.

12.2 As in 12.1,

$$M^G = \{m: m \in M \text{ and } gm = m \text{ for every } g \in G\}.$$

We observe that for any $m \in M$, the element $\sum_{g \in G} gm$ is in M^G. For, if

$h \in G$, we have

$$h \sum_{g \in G} gm = \sum_{g \in G} (hg)m.$$

Since hg ranges over all of G as g ranges over G, the last summation equals $\sum_{g \in G} gm$. Define the map $T\colon M \to M^G$ by $Tm = \sum_{g \in G} gm$ T is called the *trace* map. We note that in Galois theory since the abelian group M is the multiplicative group $K^\times$ of the field K, the trace Tm for an element $m \in K^\times$ is written multiplicatively $\prod_{g \in G} gm$ (the norm from K to k of m).

In general if $S \leqq G$, define

$$T_{S,G}\colon M^S \to M^G$$

as follows: If $G = \dot{\bigcup}_{\iota \in I} g_\iota S$ (disjoint union) where $g_\iota S$ is a left coset of S, define $T_{S,G}m = \sum_{\iota \in I} g_\iota m$; $T_{S,G}$ is well defined for if $G = \dot{\bigcup}_{\iota \in I} g_\iota s_\iota S$, then

$$T_{S,G}m = \sum_{\iota \in I} g_\iota s_\iota m = \sum_{\iota \in I} g_\iota m.$$

$T_{S,G}m \in M^G$ for every $m \in M^S$. For, if $h \in G$, from $G = \dot{\bigcup}_{\iota \in I} g_\iota S$ we have $G = hG = \dot{\bigcup}_{\iota \in I} hg_\iota S$. Thus

$$T_{S,G}m = \sum_{\iota \in I} hg_\iota m = hT_{S,G}m.$$

$T_{S,G}$ is called the *trace from S to G*. We note that

$$T = T_{1,G}.$$

Finally if

$$S \leqq U \leqq G, \qquad G = \dot{\bigcup_{\iota \in I}} g_\iota U, \qquad U = \dot{\bigcup_{\kappa \in K}} u_\kappa S,$$

(both disjoint unions), then from

$$G = \dot{\bigcup_{\iota \in I, \kappa \in K}} g_\iota u_\kappa S$$

we have

$$T_{S,G}m = \sum_{\iota,\kappa} g_\iota u_\kappa m = \sum_{\iota \in I} g_\iota \Big(\sum_{\kappa \in K} u_\kappa m\Big) = T_{U,G}T_{S,U}m.$$

Therefore, $S \leqq U \leqq G$ implies

$$T_{S,G} = T_{U,G}T_{S,U}.$$

12.3 Any G-modules M, N are $\mathbf{Z}$-modules as well. We will denote $\mathrm{Hom}_{\mathbf{Z}}(M, N)$ by $\mathrm{Hom}(M, N)$, and $\mathrm{Hom}_{\mathbf{Z}G}(M, N)$ by $\mathrm{Hom}_G(M, N)$. We define a structure of G-module on $\mathrm{Hom}(M, N)$ as follows: For lack of unambiguous symbol we denote the (left) action of $g \in G$ on $\varphi \in \mathrm{Hom}(M, N)$ by φ^g and define $\varphi^g(m) = g\varphi(g^{-1}m)$. The conditions (i)–(iii) of 11.2 are satisfied:

(i) $\varphi^1 = \varphi$
(ii) $(\varphi + \psi)^g = g(\varphi + \psi)g^{-1} = g\varphi g^{-1} + g\psi g^{-1} = \varphi^g + \psi^g$
(iii) $(\varphi^h)^g = g(h\varphi h^{-1})g^{-1} = (gh)\varphi(gh)^{-1} = \varphi^{gh}$.

Thus $\mathrm{Hom}(M, N)$ is a G-module.

If $\varphi \in (\mathrm{Hom}(M, N))^G$, then $g\varphi g^{-1} = \varphi$ for every $g \in G$; i.e., $g\varphi(m) = \varphi(gm)$. Thus

$$\sum_{g\in G} z_g g\varphi(m) = \varphi\Big(\sum_{g\in G} z_g g(m)\Big),$$

and therefore φ is a G-homomorphism. Conversely, if $\varphi \in \mathrm{Hom}_G(M, N)$, then G acts trivially on φ. Hence

$$(\mathrm{Hom}(M, N))^G = \mathrm{Hom}_G(M, N).$$

Similarly if M and N are G-modules, we have $(\mathrm{Hom}(M, N))^S = \mathrm{Hom}_S(M, N)$ for every $S \leq G$. Let $G = \dot{\bigcup}_{\iota\in I} g_\iota S$ and $\varphi \in \mathrm{Hom}_S(M, N)$, then $T_{S,G}$: $\mathrm{Hom}_S(M, N) \to \mathrm{Hom}_G(M, N)$ is given by

$$T_{S,G}\varphi = \sum_{\iota\in I} \varphi^{g_\iota} = \sum_{\iota\in I} g_\iota \varphi g_\iota^{-1}.$$

If, moreover, $\psi \in \mathrm{Hom}_G(N, P)$, then $\psi\varphi\colon M \to P$ may be viewed as an S-homomorphism, and hence

$$T_{S,G}\psi\varphi = \sum_{\iota\in I} g_\iota \psi g_\iota^{-1} g_\iota \varphi g_\iota^{-1} = \psi \sum_{\iota\in I} \varphi^{g_\iota} = \psi \cdot T_{S,G}\varphi.$$

Similarly if $\chi\colon L \to M$ is a G-homomorphism, $T_{S,G}\varphi\chi = (T_{S,G}\varphi)\chi$. In particular, if n is the order of the group G and $\varphi \in \mathrm{Hom}_G(M, N)$, then $T\varphi = n\varphi$.

12.4 Let M be a G-module. View $\mathbf{Z}$ as a G-module with trivial action, i.e., $gz = z$ for every $g \in G$ and $z \in \mathbf{Z}$. Define the map α: $\mathrm{Hom}(\mathbf{Z}, M) \to M$ by $\alpha(\varphi) = \varphi(1)$, and β: $M \to \mathrm{Hom}(\mathbf{Z}, M)$ by $(\beta(m))(1) = m$. Then α and β are G-homomorphisms, for,

$$\alpha(\varphi^g) = \varphi^g(1) = g\varphi(g^{-1}1) = g\varphi(1) = g\alpha(\varphi)$$

and

$$(\beta(gm))(1) = gm = g(\beta(m))(1).$$

Moreover

$$((\beta\alpha)\varphi)(1) = (\beta(\alpha(\varphi)))(1) = \alpha(\varphi) = \varphi(1)$$

and

$$(\alpha\beta)(m) = \alpha(\beta(m)) = (\beta(m))(1) = m.$$

Therefore we may identify $\operatorname{Hom}(\mathbf{Z}, M)$ with M by means of the G-isomorphism α. The restriction

$$\alpha|_{\operatorname{Hom}_G(\mathbf{Z},M)}: \operatorname{Hom}_G(\mathbf{Z}, M) \to M^G$$

is also an isomorphism. To show that the codomain of $\alpha|_{\operatorname{Hom}_G(\mathbf{Z},M)}$ is M^G, let $\varphi \in \operatorname{Hom}_G(\mathbf{Z}, M)$. Then

$$g\alpha|_{\operatorname{Hom}_G(\mathbf{Z},M)}(\varphi) = \varphi^g(1) = \varphi(1) = \alpha|_{\operatorname{Hom}_G(\mathbf{Z},M)}(\varphi).$$

13. *G*-Induced Modules

In this section we will use the following elementary construction: If X is a G-free module with a basis $\{x_\kappa\}_{\kappa\in K}$, $S \leqq G$, and $G = \dot{\bigcup}_{\iota\in I} Sg_\iota$, then $\{g_\iota x_\kappa\}_{\iota\in I,\kappa\in K}$ is an S-basis of X viewed as an S-free module.

13.1 A G-module M is called G-*induced* (or G-*regular*) if M contains a $\mathbf{Z}$-module N such that $M = \sum_{g\in G} gN$ (direct). If $S \leqq G$ and M is G-induced, then M is S-induced. In fact, let $G = \dot{\bigcup}_{\kappa\in K} Sg_\kappa$ and $M = \sum_{g\in G} gN$ for some $\mathbf{Z}$-module N. Then

$$M = \sum_{s\in S} sN_1 \qquad \text{where} \quad N_1 = \sum_{\kappa\in K} g_\kappa N.$$

A G-free module M is G-induced; for if $\{x_\iota\}_{\iota\in I}$ is a G-basis of M, then

$$M = \sum_{g\in G} gN \qquad \text{where} \quad N = \sum_{\iota\in I} \mathbf{Z}x_\iota.$$

13.2 *Let $M = \sum_{g \in G} gN$ be G-induced. Then there exists $\pi \in \mathrm{Hom}(M, M)$ such that $1_M = T\pi$ where T is the trace map.*

Let $M \xrightarrow{\pi} M = M \xrightarrow{p_1} 1N \xrightarrow{i_1} M$ where p_1 is the projection map from the direct sum $\sum_{g \in G} gN$ to the summand $1N$ and i_1 is the injection of the summand $1N$ in the direct sum $\sum_{g \in G} gN$. Then if $m = \sum_{g \in G} gn_g \in M$, we have $\pi m = n_1 \in M$. For any $h \in G$,

$$\pi^h m = h\pi(h^{-1}m) = h\pi(\sum_{g \in G} h^{-1}gn_g) = hn_h.$$

Therefore

$$T\pi m = \sum_{h \in G} \pi^h m = \sum_{h \in G} hn_h = m.$$

13.3 *If M is G-induced and L is a G-module, then* $\mathrm{Hom}(M, L)$ *and $M \otimes L$ are G-induced.*

Construct $\pi \colon M \to M$ as in 13.2 and let $N_1 = \mathrm{Hom}(M, L)\cdot\pi \leq \mathrm{Hom}(M, L)$. Then $\mathrm{Hom}(M, L) = \sum_{g \in G} gN_1$. For, if $\varphi \in \mathrm{Hom}(M, L)$, then

$$\begin{aligned} \varphi &= \varphi 1_M \\ &= \varphi T\pi \qquad \text{by 13.2} \\ &= \sum_{g \in G} \varphi\pi^g \\ &= \sum_{g \in G} (\varphi^{g^{-1}})^g \pi^g. \end{aligned}$$

Moreover $\sum_{g \in G} gN_1$ is direct since the representation $\varphi = \sum_{g \in G} (\varphi^{g^{-1}})^g \pi^g$ is unique.

To show $M \otimes L$ is G-induced we note that if $M = \sum_{g \in G} gN$, then

$$M \otimes L = (\sum_{g \in G} gN) \otimes L = \sum_{g \in G} (g \otimes g)(N \otimes L)$$

since $gL = L$. Moreover $\sum_{g \in G} (g \otimes g)(N \otimes L)$ is direct by 3.2.

14. Complete Resolutions

14.1 Associate with the finite group G any diagram

$$(1)\qquad \cdots \xleftarrow{\partial_{-2}} X_{-2} \xleftarrow{\partial_{-1}} X_{-1} \xleftarrow{\partial_0} X_0 \xleftarrow{\partial_1} X_1 \xleftarrow{\partial_2} \cdots$$

$$\begin{array}{ccc} X_{-1} & & X_0 \\ \mu \nwarrow & & \swarrow \varepsilon \\ & \mathbf{Z} & \\ \swarrow & & \nwarrow \\ 0 & & 0 \end{array}$$

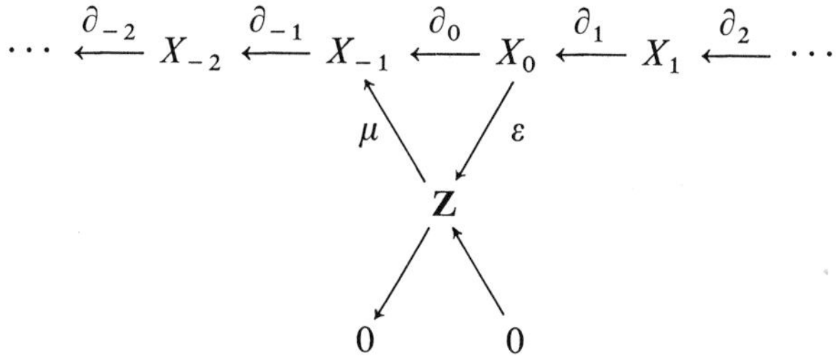

where:

(i) The row is an acyclic sequence of homomorphisms of finitely generated G-free modules.

(ii) $\partial_0 = \mu\varepsilon$.

(iii) The sequence $X_{-2} \xleftarrow{\partial_{-1}} X_{-1} \xleftarrow{\mu} \mathbf{Z} \leftarrow 0$ is exact.

(iv) The sequence $0 \leftarrow \mathbf{Z} \xleftarrow{\varepsilon} X_0 \xleftarrow{\partial_1} X_1$ is exact.

ε is called the *augmentation map* and μ is called the *coaugmentation map*. The diagram (1) is called a *complete resolution* of G if (i)–(iv) are satisfied, and is denoted by X_*.

The modules X_r are $\mathbf{Z}$-free since they are G-free. Thus by the acyclicity of the row in (1) and 9.4 it follows that there exists a family $\{D_r : X_r \to X_{r+1}\}_{r \in \mathbf{Z}}$ of $\mathbf{Z}$-homomorphisms such that $1_r = D_{r-1}\partial_r + \partial_{r+1}D_r$ for every $r \in \mathbf{Z}$ where 1_r denotes the identity homomorphism $1_{X_r}: X_r \to X_r$.

14.2 *Example.* Suppose G is a cyclic group of order n generated by g. Let $\mathbf{Z}G$ be, as always, the group ring of G. The diagram

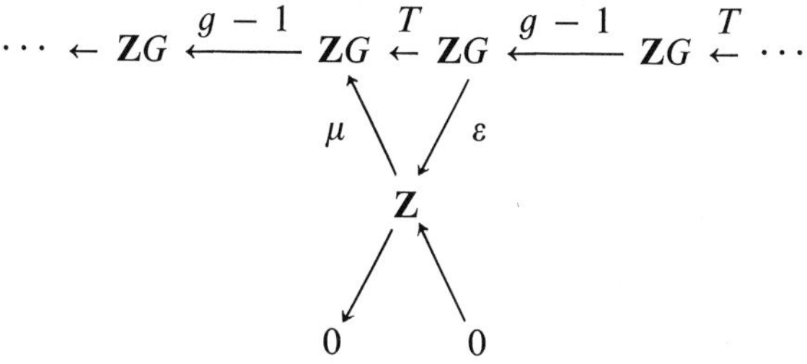

$$\cdots \leftarrow \mathbf{Z}G \xleftarrow{g-1} \mathbf{Z}G \xleftarrow{T} \mathbf{Z}G \xleftarrow{g-1} \mathbf{Z}G \xleftarrow{T} \cdots$$

$$\begin{array}{ccc} \mathbf{Z}G & & \mathbf{Z}G \\ \mu \nwarrow & & \swarrow \varepsilon \\ & \mathbf{Z} & \\ \swarrow & & \nwarrow \\ 0 & & 0 \end{array}$$

of G-homomorphisms, where

$$\partial_{\text{odd}} = g - 1, \qquad \partial_{\text{even}} = T = \sum_{i=0}^{n-1} g^i$$

$$\varepsilon(g^i) = 1, \qquad \mu(1) = T,$$

is a complete resolution of the cyclic group G. We will demonstrate this in (1)–(6) below.

(1) μ has kernel zero.
(2) ε is onto.
(3) $T = \mu\varepsilon$, for, $\mu\varepsilon(g^i) = \mu(1) = T = Tg^i$.

(4) $$(g - 1)T = (g - 1)\sum_{i=0}^{n-1} g^i = \sum_{i=0}^{n-1} g^{i+1} - \sum_{i=0}^{n-1} g^i = 0.$$

Similarly, $T(g - 1) = 0$.
(5) Suppose

$$\sum_{i=0}^{n-1} z_i g^i \in \mathbf{Z}G \qquad \text{and} \qquad (g - 1)\sum_{i=0}^{n-1} z_i g^i = 0,$$

then

$$(g - 1)\sum_{i=0}^{n-1} z_i g^i = \sum_{i=0}^{n-1} (z_{i-1} - z_i)g^i = 0.$$

Hence $z_{i-1} = z_i = z$. Therefore

$$\sum_{i=0}^{n-1} z_i g^i = z\sum_{i=0}^{n-1} g^i = Tz.$$

(6) Suppose $T\sum_{i=0}^{n-1} z_i g^i = 0$, then

$$0 = T\sum_{i=0}^{n-1} z_i g^i = \mu\sum_{i=0}^{n-1} z_i$$

by (3). Therefore, by (1), $\sum_{i=0}^{n-1} z_i = 0$. Hence

$$\sum_{i=0}^{n-1} z_i g^i = \sum_{i=0}^{n-1} z_i(g^i - 1)$$

$$= z_1(g - 1) + \sum_{i=2}^{n-1} z_i(g - 1)(g^{i-1} + g^{i-2} + \cdots + 1).$$

Thus

$$(g - 1)(z_1 + \sum_{i=2}^{n-1} z_i(g^{i-1} + g^{i-2} + \cdots + 1)) = \sum_{i=0}^{n-1} z_i g^i.$$

We will return to this example in 23.

14.3 Given the complete resolution (1) of G in 14.1 and the G-module M, consider the cochain complex of **Z**-homomorphisms

$$\text{(2)} \qquad \cdots \to M^{-2} \xrightarrow{\delta^{-2}} M^{-1} \xrightarrow{\delta^{-1}} M^0 \xrightarrow{\delta^0} M^1 \to \cdots$$

where $M^r = \operatorname{Hom}_G(X_r, M)$ and $\delta^r = \operatorname{Hom}_G(\partial_{r+1}, 1_M)$. Recall that if $t \in \operatorname{Hom}_G(X_r, M)$, then

$$\delta^r t = \operatorname{Hom}_G(\partial_{r+1}, 1_M) = 1_M t \partial_{r+1}$$

and $\delta^r t$ is called the rth coboundary of t. Equation (2) is a cochain complex since

$$\begin{aligned}\delta^{r+1}\delta^r &= \operatorname{Hom}_G(\partial_{r+2}, 1_M)\operatorname{Hom}_G(\partial_{r+1}, 1_M) = \operatorname{Hom}_G(\partial_{r+1}\partial_{r+2}, 1_M) \\ &= 0.\end{aligned}$$

The cohomology groups of the cochain complex (2) are called the *cohomology groups of the finite group G with coefficients in the G-module M*, and are denoted by $H^r(G, M)$. The abelian group $H^r(G, M)$ is also called the rth *Tate group* of G with coefficients in M. We point out at once that our construction of the groups $H^r(G, M)$ depends on

(i) the existence of complete resolutions of G;
(ii) the choice of a complete resolution of G.

We will show in Sections 16 and 17 that complete resolutions of G exist and that $H^r(G, M)$ is independent of the choice of the resolution. Before that we will study some of the properties of $H^r(G, M)$ and consider some examples.

15. The Dependence of $H^r(G,M)$ on the G-Module M

15.1 Let $\varphi: M \to N$ be a G-homomorphism. Suppose a complete resolution X_* (as in 14.1) of G is given. Then

$$\operatorname{Hom}_G(1_{X_r}, \varphi): \operatorname{Hom}_G(X_r, M) \to \operatorname{Hom}_G(X_r, N)$$

is a **Z**-homomorphism defined by $\operatorname{Hom}_G(1_{X_r}, \varphi)t = \varphi t$.

As an immediate consequence we have the commutativity of the diagram

(1)

$$\begin{array}{ccccccc}
\cdots \to M^{r-1} & \xrightarrow{\delta^{r-1}} & M^r & \xrightarrow{\delta^r} & M^{r+1} \to \cdots \\
\Big\downarrow \mathrm{Hom}_G(1_{X_{r-1}}, \varphi) & & \Big\downarrow \mathrm{Hom}_G(1_{X_r}, \varphi) & & \Big\downarrow \mathrm{Hom}_G(1_{X_{r+1}}, \varphi) \\
\cdots \to N^{r-1} & \xrightarrow{\delta^{r-1}} & N^r & \xrightarrow{\delta^r} & N^{r+1} \to \cdots
\end{array}$$

where $N^r = \mathrm{Hom}_G(X_r, N)$. Thus $\mathrm{Hom}_G(1_{X*}, \varphi)$ is a morphism of cochain complexes. Moreover, by 8.2, the morphism $\mathrm{Hom}_G(1_{X*}, \varphi)$ induces the $\mathbf{Z}$-homomorphism $\overline{(1, \varphi)}^r: H^r(G, M) \to H^r(G, N)$ for every r, where the image under $\overline{(1, \varphi)}^r$ of the cocycle class containing t is the cocycle class containing $\mathrm{Hom}_G(1_{X_r}, \varphi)t = \varphi t$.

15.2 Let

$$0 \to L \xrightarrow{i} M \xrightarrow{j} N \to 0$$

be an exact sequence of G-homomorphisms. Applying $\mathrm{Hom}_G(X_*, \ )$ to this sequence we get the commutative diagram

$$\begin{array}{ccccc}
\vdots & & \vdots & & \vdots \\
\downarrow & & \downarrow & & \downarrow \\
0 \to L^{r-1} & \xrightarrow{\mathrm{Hom}_G(1_{X_{r-1}}, i)} & M^{r-1} & \xrightarrow{\mathrm{Hom}_G(1_{X_{r-1}}, j)} & N^{r-1} \to 0 \\
\Big\downarrow \delta^{r-1} & & \Big\downarrow \delta^{r-1} & & \Big\downarrow \delta^{r-1} \\
0 \to L^r & \xrightarrow{\mathrm{Hom}_G(1_{X_r}, i)} & M^r & \xrightarrow{\mathrm{Hom}_G(1_{X_r}, j)} & N^r \to 0 \\
\Big\downarrow \delta^r & & \Big\downarrow \delta^r & & \Big\downarrow \delta^r \\
0 \to L^{r+1} & \xrightarrow{\mathrm{Hom}_G(1_{X_{r+1}}, i)} & M^{r+1} & \xrightarrow{\mathrm{Hom}_G(1_{X_{r+1}}, j)} & N^{r+1} \to 0 \\
\downarrow & & \downarrow & & \downarrow \\
\vdots & & \vdots & & \vdots
\end{array}$$

where the columns are cochain complexes and the rows are exact since

X_r is a G-free module for every r (see 7.7). By 8.4, we have the exact sequence

$$\cdots \to H^{r-1}(G, L) \to H^{r-1}(G, M) \to H^{r-1}(G, N) \to H^r(G, L) \to \cdots .$$

15.3 Let $\varphi: M \to N$ be a G-homomorphism such that $\varphi = T\psi$ for some $\psi \in \mathrm{Hom}(M, N)$ where T is the trace map. If X_* is a complete resolution of G, we have the commutative diagram

$$\begin{array}{ccccccc}
\cdots \to M^{r-1} & \xrightarrow{\delta^{r-1}} & M^r & \xrightarrow{\delta^r} & M^{r+1} \to \cdots \\
\downarrow \mathrm{Hom}_G(1_{X_{r-1}}, \varphi) & & \downarrow \mathrm{Hom}_G(1_{X_r}, \varphi) & & \downarrow \mathrm{Hom}_G(1_{X_{r+1}}, \varphi) \\
\cdots \to N^{r-1} & \xrightarrow{\delta^{r-1}} & N^r & \xrightarrow{\delta^r} & N^{r+1} \to \cdots
\end{array}$$

where $M^r = \mathrm{Hom}_G(X_r, M)$ and $N^r = \mathrm{Hom}_G(X_r, N)$. Let t be an r-cocycle in M^r, i.e., $t \in \mathrm{Hom}_G(X_r, M)$ such that $\delta^r t = t\partial_{r+1} = 0$. If we view X_* as an acyclic complex of homomorphisms of $\mathbf{Z}$-free modules, by 14.1 we have

$$\psi t = \psi t 1_{X_r} = \psi t(D_{r-1}\partial_r + \partial_{r+1}D_r) = \psi t D_{r-1}\partial_r.$$

Applying the trace map we get

$$\begin{aligned}
T(\psi t) &= (T\psi)t \qquad \text{since} \quad t \text{ is a } G\text{-homomorphism} \\
&= \varphi t \\
&= \mathrm{Hom}_G(1_{X_r}, \varphi)t
\end{aligned}$$

on one hand, and

$$\begin{aligned}
T(\psi t D_{r-1}\partial_r) &= (T(\psi t D_{r-1}))\partial_r \qquad \text{since} \qquad \partial_r \text{ is a } G\text{-homomorphism} \\
&= \delta^{r-1} T(\psi t D_{r-1})
\end{aligned}$$

on the other hand. Summarizing, we have the following:

If the G-homomorphism $\varphi: M \to N$ is such that $\varphi = T\psi$ for some $\psi \in \mathrm{Hom}(M, N)$, where T is the trace map, then the induced $\mathbf{Z}$-homomorphism $\overline{(1, \varphi)}^r: H^r(G, M) \to H^r(G, N)$ is trivial.

15.4 If in particular M is G-induced, then, by 13.2, $1_M = T\pi$ for some $\pi \in \mathrm{Hom}(M, M)$. Therefore

$$\overline{(1, 1_M)^r}: H^r(G, M) \to H^r(G, M)$$

is simultaneously the zero homomorphism and the identity homomorphism. Hence:

If M is G-induced, then $H^r(G, M) = 0$ for every r.

15.5 Let M be a G-module, and n the order of the group G. The G-homomorphism $n1_M: M \to M$ defined by $n1_M(m) = nm$ is the trace of 1_M. For,

$$T(1_M) = \sum_{g \in G} g 1_M g^{-1} = \sum_{g \in G} (gg^{-1}) 1_M = n1_M.$$

Therefore, by 15.3, $\overline{(1, n1_M)^r}: H^r(G, M) \to H^r(G, M)$ is the zero map. But, by 9.1,

$$\overline{(1, n1_M)^r} = \overline{(1, 1_M)^r} + \cdots + \overline{(1, 1_M)^r} = n1_{H^r(G,M)}.$$

We have shown:
The order of each element of $H^r(G, M)$ is a divisor of the order of G.

15.6 We consider now the case where M is finitely generated, say by $m_1, \ldots, m_t$ over $\mathbf{Z}G$. Let X_* be a complete resolution of G and let $x_1, \ldots, x_s$ be a basis of X_r over $\mathbf{Z}G$. Let $\varphi_{\iota\kappa}: X_r \to M$ be the G-homomorphism that sends x_ι to m_κ and the other x_v to zero. Adding G-multiples of such G-homomorphisms $\varphi_{\iota\kappa}$ where the summation ranges over κ we can construct a G-homomorphism φ_ι that sends x_ι to a given $m \in M$, and the other x_v to zero. By 1.2, any G-homomorphism $\varphi: X_r \to M$ is of the form $\sum_\iota \varphi_\iota$. Therefore if M is a finitely generated G-module, $M^r = \mathrm{Hom}_G(X_r, M)$ is finitely generated. It follows that the subgroup of the r-cocycles is finitely generated. Thus $H^r(G, M)$ is finitely generated. Moreover since, by 15.5, every element of $H^r(G, M)$ has finite order, we conclude that:

If M is a finitely generated G-module, then $H^r(G, M)$ is finite.

In particular, since the G-module $\mathbf{Z}$ is generated by 1 over $\mathbf{Z}G$ (the action of G on $\mathbf{Z}$ is trivial), $H^r(G, \mathbf{Z})$ is finite.

15.7 *Example.* Let K be a finite normal separable extension of a field k. Let G be the Galois group of K over k; G acts on K and its action restricted to k is trivial. View K^+, the additive group of K, as a G-module by suppressing the multiplicative structure of K. We wish to compute $H^r(G, K^+)$. For $a \in K$ define the $\mathbf{Z}$-homomorphism $\varphi_a: K^+ \to K^+$ by $\varphi_a(x) = ax$. View $\operatorname{Hom}(K^+, K^+)$ as a G-module (see 12.3), then for $g \in G$,

$$\varphi_a^g(x) = g(\varphi_a g^{-1}x) = g(a \cdot g^{-1}x) = ga \cdot x.$$

Thus $\varphi_a^g = \varphi_{ga}$. Since $\varphi_a + \varphi_b = \varphi_{a+b}$, we have

$$T\varphi_a = \sum_{g \in G} \varphi_a^g = \varphi_{Ta}.$$

We claim there exists $a \in K$ such that $Ta = 1 \in k$. In fact, if the characteristic of K is zero. $T(1/n) = 1$ where n is the order of G. In case of arbitrary characteristic we show in Chapter 10, using a well-known theorem of Artin, that there exists $b \in K$ such that $T(b) = \beta \neq 0$. [This can also be verified by showing that $K = k(x)$ for some primitive element $x \in K$, and that the discriminant $d_{1,x,\ldots,x^{n-1}} \neq 0$. Moreover this holds if and only if there exists some $b \in K$ such that $Tb \neq 0$.] Of course, $\beta \in k$ and $T(b/\beta) = 1$. Setting $a = b/\beta$, we have $T\varphi_a = \varphi_{Ta} = \varphi_1 = 1_{K^+}$. Hence, by 15.3, $H^r(G, K^+) = 0$ for every r.

15.8 If $n1_M: M \to M$ is an isomorphism, then M is called *uniquely divisible* by n. If this is the case, then the map $(1/n)1_M: M \to M$ is the inverse of the map $n1_M$. Hence

$$1_{H^r(G,M)} = (\overline{1, 1_M})^r = (\overline{1, n1_M})^r \cdot (\overline{1,(1/n)1_M})^r.$$

If $n = |G|$, by 15.5, $(\overline{1, n1_M})^r = 0$. We have shown that:

If the G-module M is uniquely divisible by the order of G, then $H^r(G, M) = 0$ for every r.

15.9 *Example.* Let K be a field whose characteristic does not divide the order of the group G. Let V be any vector space over K. Let G act on

K in any way. Since V is uniquely divisible by the order of G, we have $H^r(G, V) = 0$ for every r. Thus if the characteristic of K is zero, we have $H^r(G, V) = 0$ for any finite group G and any action of G on K.

15.10 *Example.* Let $\mathbf{Q}$ be the rational numbers. Consider the exact sequence of trivial G-modules

$$0 \to \mathbf{Z} \to \mathbf{Q} \to \mathbf{Q}/\mathbf{Z} \to 0.$$

By 15.2, we have the exact sequence

$$\cdots \to H^r(G, \mathbf{Q}) \to H^r(G, \mathbf{Q}/\mathbf{Z}) \xrightarrow{\bar{\delta}} H^{r+1}(G, \mathbf{Z}) \to H^{r+1}(G, \mathbf{Q}) \to \cdots.$$

Since, by 15.8, $H^r(G, \mathbf{Q}) = 0$ for every r, we have

$$H^r(G, \mathbf{Q}/\mathbf{Z}) \overset{\bar{\delta}}{\cong} H^{r+1}(G, \mathbf{Z}).$$

16. Existence of Complete Resolutions

16.1 *The bar resolution of the G-module* $\mathbf{Z}$. Let X_0 be the free G-module generated by the symbol [ ]. Thus X_0 may be viewed as a free $\mathbf{Z}$-module freely generated by the symbols $g[\ ]$, $g \in G$. Define the G-homomorphism $\varepsilon: X_0 \to \mathbf{Z}$ by $\varepsilon[\ ] = 1$. Let X_r be the free G-module freely generated by the symbols $[g_1, \ldots, g_r]$ where $g_\iota \in G$. Thus X_r is a free $\mathbf{Z}$-module freely generated by the symbols $g[g_1, \ldots, g_r]$ where $g, g_\iota \in G$. Define the G-homomorphisms

$$\varepsilon_r^\iota: X_r \to X_{r-1} \qquad \text{for} \quad \iota = 0, 1, \ldots, r$$

called *faces* by

$$\begin{aligned}
\varepsilon_r^0[g_1, \ldots, g_r] &= g_1[g_2, \ldots, g_r] \\
\varepsilon_r^\iota[g_1, \ldots, g_r] &= [g_1, \ldots, g_\iota g_{\iota+1}, \ldots, g_r] \qquad \text{for} \quad 1 < \iota < r \\
\varepsilon_r^r[g_1, \ldots, g_r] &= [g_1, \ldots, g_{r-1}].
\end{aligned}$$

Define the G-homomorphism

$$\partial_r = \sum_{\iota=0}^{r} (-1)^\iota \varepsilon_r^\iota.$$

Define the **Z**-homomorphisms $E: \mathbf{Z} \to X_0$ by $E(1) = [\ ]$, and $D_r: X_r \to X_{r+1}$ by

$$D_r(g[g_1, \ldots, g_r]) = [g, g_1, \ldots, g_r].$$

Consider the diagram

$$(1) \qquad 0 \leftarrow \mathbf{Z} \underset{E}{\overset{\varepsilon}{\leftrightarrows}} X_0 \underset{D_0}{\overset{\partial_1}{\leftrightarrows}} X_1 \underset{D_1}{\overset{\partial_2}{\leftrightarrows}} \cdots$$

viewed as a diagram of **Z**-homomorphisms. We observe that

$$(2) \qquad \varepsilon E(1) = 1.$$

$$(3) \qquad E\varepsilon + \partial_1 D_0 = 1_{X_0},$$

for,

$$(E\varepsilon)g[\ ] = E(1) = [\ ], \qquad (\varepsilon_1^0 D_0)g[\ ] = g[\ ], \qquad \text{and}$$
$$(\varepsilon_1^1 D_0)g[\ ] = [\ ].$$

$$(4) \qquad \partial_{r+1}D_r + D_{r-1}\partial_r = 1_{X_r},$$

for,

$$\partial_{r+1}D_r + D_{r-1}\partial_r = \varepsilon_{r+1}^0 D_r + \sum_{\iota=0}^{r} \{(-1)^{\iota+1}\varepsilon_{r+1}^{\iota+1}D_r + (-1)^{\iota}D_{r-1}\varepsilon_r^{\iota},$$

and

$$\varepsilon_{r+1}^0 D_r g[g_1, \ldots, g_r] = \varepsilon_{r+1}^0[g, g_1, \ldots, g_r] = g[g_1, \ldots, g_r]$$

whereas

$$\varepsilon_{r+1}^{\iota+1}D_r = D_{r-1}\varepsilon_r^{\iota} \qquad \text{for} \quad \iota = 0, \ldots, r.$$

The last identity holds since:

for $\iota = 0$,

$$\varepsilon_{r+1}^1 D_r g[g_1, \ldots, g_r] = [gg_1, g_2, \ldots, g_r] = D_{r-1}\varepsilon_r^0 g[g_1, \ldots, g_r],$$

for $0 < \iota < r$,

$$\varepsilon_{r+1}^{\iota+1}D_r g[g_1, \ldots, g_r] = [g, g_1, \ldots, g_\iota g_{\iota+1}, \ldots, g_r]$$
$$= D_{r-1}\varepsilon_r^{\iota} g[g_1, \ldots, g_r],$$

for $\iota = r + 1$,

$$\varepsilon_{r+1}^{r+1} D_r g[g_1, \ldots, g_r] = [g, g_1, \ldots, g_{r-1}] = D_{r-1} \varepsilon_r^r g[g_1, \ldots, g_r].$$

We claim the sequence

$$(5) \qquad 0 \leftarrow \mathbf{Z} \overset{\varepsilon}{\leftarrow} X_0 \overset{\partial_1}{\leftarrow} X_1 \leftarrow \cdots$$

is a complex of G-homomorphisms. This is verified in (i)–(iii) below.

(i) By (3) above, $\varepsilon = \varepsilon 1_{X_0} = \varepsilon E \varepsilon + \varepsilon \partial_1 D_0$. But $\varepsilon E = 1_{\mathbf{Z}}$, therefore $\varepsilon\, \partial_1 D_0 = 0$. Since D_0 has in its image all the G-generators of X_1, $\varepsilon\, \partial_1 = 0$ on the G-generators of X_1. Thus $\varepsilon\, \partial_1 = 0$.

(ii) Multiply the identity (3) on the right by ∂_1. By (i), $\varepsilon\, \partial_1 = 0$, therefore

$$(6) \qquad \partial_1 = \partial_1 D_0\, \partial_1.$$

Since $1_{X_1} = D_0 \partial_1 + \partial_2 D_1$ implies $\partial_1 = \partial_1 D_0 \partial_1 + \partial_1 \partial_2 D_1$, the substitution of (6) in the last identity results in $\partial_1 \partial_2 D_1 = 0$. We observe that the image of D_1 contains all the G-generators of X_2. Therefore $\partial_1 \partial_2$ is zero on all G-generators of X_2. Thus $\partial_1 \partial_2 = 0$.

(iii) By induction suppose $\partial_{r-1} \partial_r = 0$. Multiply

$$1_{X_{r-1}} = D_{r-2} \partial_{r-1} + \partial_r D_{r-1}$$

on the right by ∂_r to obtain the identity

$$(7) \qquad \partial_r = \partial_r D_{r-1} \partial_r.$$

Multiply $1_{X_r} = D_{r-1} \partial_r + \partial_{r+1} D_r$ on the left by ∂_r and substitute (7) in the resulting identity to get $\partial_r\, \partial_{r+1} D_r = 0$. Since the image of D_r contains all the G-generators of X_{r+1}, we conclude that $\partial_r \partial_{r+1} = 0$. Summarizing, we have shown that:

The sequence (5) *of G-homomorphisms is a complex. Moreover, when it is viewed as a complex of* $\mathbf{Z}$*-homomorphisms, the identity morphism of this complex is homotopic to zero* [*this follows from* (2), (3), (4)]. *Therefore by 9.3 the complex* (5) *viewed as a complex of* $\mathbf{Z}$*-homomorphisms is acyclic. Hence* (5) *viewed as a complex of G-homomorphisms is acyclic.* The acyclic (i.e., exact) complex (5) is called the *bar resolution* of the G-module $\mathbf{Z}$.

16.2 *In this section let*

$$0 \leftarrow \mathbf{Z} \stackrel{\varepsilon}{\leftarrow} X_0 \stackrel{\partial_1}{\leftarrow} X_1 \stackrel{\partial_2}{\leftarrow} \cdots \tag{8}$$

be any acyclic complex of homomorphisms of G-free modules, where the action of G on $\mathbf{Z}$ *is trivial* ($gz = z$ *for every* $z \in \mathbf{Z}$ *and* $g \in G$). *We will show (8) can be extended to a complete resolution of G.* This together with 16.1 will constitute a proof of the existence of complete resolutions of G.

The sequence (8) may be viewed as an exact sequence of homomorphisms of $\mathbf{Z}$-free modules; hence, by 9.4, there exist $\mathbf{Z}$-homomorphisms

$$0 \to \mathbf{Z} \stackrel{E}{\to} X_0 \stackrel{D_0}{\longrightarrow} X_1 \stackrel{D_1}{\longrightarrow} X_2 \to \cdots$$

such that the following identities of $\mathbf{Z}$-homomorphisms hold:

$$1_{\mathbf{Z}} = \varepsilon E \tag{9}$$

$$1_{X_0} = E\varepsilon + \partial_1 D_0 \tag{10}$$

$$1_{X_r} = D_{r-1}\partial_r + \partial_{r+1}D_r \qquad \text{for } r = 1, 2, \ldots. \tag{11}$$

Define $X_r^* = \mathrm{Hom}(X_r, \mathbf{Z})$ and $\mathbf{Z}^* = \mathrm{Hom}(\mathbf{Z}, \mathbf{Z}) = \mathbf{Z}$. Consider the sequence

$$\cdots \leftarrow X_2^* \stackrel{\partial_2^*}{\longleftarrow} X_1^* \stackrel{\partial_1^*}{\longleftarrow} X_0^* \stackrel{\varepsilon^*}{\leftarrow} \mathbf{Z}^* \leftarrow 0 \tag{12}$$

where $\varepsilon^* = \mathrm{Hom}(\varepsilon, 1_{\mathbf{Z}})$ and $\partial_r^* = \mathrm{Hom}(\partial_r, 1_{\mathbf{Z}})$. In 7.3 we have shown that $\mathrm{Hom}(\varphi + \psi, 1) = \mathrm{Hom}(\varphi, 1) + \mathrm{Hom}(\psi, 1)$, $\mathrm{Hom}(\varphi\psi, 1) = \mathrm{Hom}(\psi, 1)\mathrm{Hom}(\varphi, 1)$, and $\mathrm{Hom}(1, 1) = 1$. Therefore

$$1_{\mathbf{Z}^*} = E^*\varepsilon^* \tag{13}$$

$$1_{X_0^*} = \varepsilon^* E^* + D_0^*\partial_1^* \tag{14}$$

$$1_{X_r^*} = \partial_r^* D_{r-1}^* + D_r^*\partial_{r+1}^* \qquad \text{for } r = 1, 2, \ldots. \tag{15}$$

Moreover the sequence (12) is a complex of $\mathbf{Z}$-homomorphisms. In fact,

$$\partial_1^*\varepsilon^* = \mathrm{Hom}(\varepsilon\partial_1, 1_{\mathbf{Z}}) = 0$$

and

$$\partial_{r+1}^*\partial_r^* = \mathrm{Hom}(\partial_r\partial_{r+1}, 1_{\mathbf{Z}}) = 0 \qquad \text{for } r = 1, 2, \ldots.$$

We observe that the **Z**-modules X_r^* in (12) can be made into G-modules by 12.3. We claim ∂_r^* and ε^* are G-homomorphisms, for, if $\varphi \in X_r^* = \mathrm{Hom}(X_r, \mathbf{Z})$, then

$$\begin{aligned} \partial^{*g}\varphi &= \mathrm{Hom}(\partial_{r+1}, 1)^g\varphi = g\ \mathrm{Hom}(\partial_{r+1}, 1)g^{-1}\ \varphi \\ &= gg^{-1}\varphi\ \partial_{r+1} \\ &= \partial^*\varphi. \end{aligned}$$

Similarly, $\varepsilon^{*g}z = \varepsilon^* z$ for every $z \in \mathbf{Z}^*$. Finally X_r^* is G-free and finitely generated. This is a special case of 15.6 where the G-module M is replaced by the trivial G-module **Z**. The acyclicity of (12) now follows by a theorem obtained by reversing all the arrows in 9.3.

We now glue together the complexes (8) and (12) in the diagram

(16)
$$\begin{array}{ccccccccccc} & \partial_2^* & & \partial_1^* & & \partial_0 & & \partial_1 & & \partial_2 & \\ \cdots & \leftrightarrows & X_1^* & \leftrightarrows & X_0^* & \leftrightarrows & X_0 & \leftrightarrows & X_1 & \leftrightarrows & \cdots \\ & D_1^* & & D_0^* & \nwarrow\!\searrow & D_{-1} & \nearrow\!\swarrow & D_0 & & D_1 & \\ & & & & E^* & \varepsilon^* \quad \varepsilon & E & & & & \\ & & & & & \mathbf{Z} & & & & & \\ & & & & \swarrow & & \nwarrow & & & & \\ & & & & 0 & & 0 & & & & \end{array}$$

where ∂_0 is defined by $\partial_0 = \varepsilon^*\varepsilon$ and D_{-1} is defined by $D_{-1} = EE^*$. We observe that $\partial_1^*\partial_0 = \partial_1^*\, \varepsilon^*\varepsilon = 0$ and $\partial_0\partial_1 = \varepsilon^*\varepsilon\, \partial_1 = 0$. Thus the sequence of G-homomorphisms

$$\cdots \overset{\partial_2^*}{\leftarrow} X_1^* \overset{\partial_1^*}{\leftarrow} X_0^* \overset{\partial_0}{\leftarrow} X_0 \overset{\partial_1}{\leftarrow} X_1 \overset{\partial_2}{\leftarrow} \cdots$$

is a complex. This complex is exact since

(17)
$$D_{-1}\partial_0 + \partial_1 D_0 = 1_{X_0}$$

and

(18)
$$D_0^*\partial_1^* + \partial_0 D_{-1} = 1_{X_0^*}.$$

The identities (17) and (18) follow from the definitions.

17. The Uniqueness of Cohomology Groups

17.1 *Dimension shifters.* Let $i: I \to \mathbf{Z}G$ be the kernel of the augmentation map $\varepsilon: \mathbf{Z}G \to \mathbf{Z}$. Then $\sum_{g \in G} z_g g \in I$ if and only if $\sum_{g \in G} z_g = 0$. Thus if $\sum_{g \in G} z_g g \in I$, then

$$\sum_{g \in G} z_g g = \sum_{\substack{g \in G \\ g \neq 1}} z_g (g - 1),$$

i.e., I is generated as a $\mathbf{Z}$-module by the family $\{g - 1\}_{g \in G, g \neq 1}$. If $\sum_{g \in G} z_g (g - 1) = 0$, then

$$\left(\sum_{\substack{g \in G \\ g \neq 1}} - z_g\right)1 + \sum_{\substack{g \in G \\ g \neq 1}} z_g g = 0.$$

Since $\mathbf{Z}G$ is freely generated by the elements of G, the identity displayed above implies $z_g = 0$ for every $g \in G$. Therefore $\{g - 1\}_{g \in G, g \neq 1}$ is a $\mathbf{Z}$-basis of I. Let $\mu: \mathbf{Z} \to \mathbf{Z}G$ be the G-homomorphism defined by $\mu(1) = T = \sum_{g \in G} g$ where the action of G on $\mathbf{Z}$ is trivial. Let $J = \mathbf{Z}G/\mu\mathbf{Z} = \mathbf{Z}G/\mathbf{Z} \cdot T$. Then we have the exact sequences

(1a) $$0 \to I \xrightarrow{i} \mathbf{Z}G \xrightarrow{\varepsilon} \mathbf{Z} \to 0$$

(1b) $$0 \to \mathbf{Z} \xrightarrow{\mu} \mathbf{Z}G \xrightarrow{j} J \to 0$$

The sequences (1a), (1b) are $\mathbf{Z}$-split; i.e., if we view I and $\mathbf{Z}G$ as $\mathbf{Z}$-modules, then iI and $\mu\mathbf{Z}$ are direct summands of $\mathbf{Z}G$. In fact, if $\sum_{g \in G} z_g g \in \mathbf{Z}G$ then

$$\sum_{g \in G} z_g g = \sum_{g \in G} z_g (g - 1) + \left(\sum_{g \in G} z_g\right)1$$

and

$$\sum_{g \in G} z_g g = z_1 T + \sum_{g \in G} (z_g - z_1) g.$$

Moreover if $\lambda: \mathbf{Z}G \to I$ is a $\mathbf{Z}$-homomorphism defined by $\lambda(\sum_{g \in G} z_g g) = \sum_{g \in G} z_g (g - 1)$, and $\rho: \mathbf{Z}G \to \mathbf{Z}$ is the $\mathbf{Z}$-homomorphism defined by $\rho(\sum_{g \in G} z_g g) = z_1$, then $\lambda i = 1_I$ and $\rho\mu = 1_{\mathbf{Z}}$.

Let M be a G-module. Viewing M as a $\mathbf{Z}$-module, the sequences

$$(2)\qquad 0 \to I \otimes M \xrightarrow{i \otimes 1_M} \mathbf{Z}G \otimes M \xrightarrow{\varepsilon \otimes 1_M} \mathbf{Z} \otimes M \to 0$$

$$0 \to \mathbf{Z} \otimes M \xrightarrow{\mu \otimes 1_M} \mathbf{Z}G \otimes M \xrightarrow{j \otimes 1_M} J \otimes M \to 0$$

$$0 \to \mathrm{Hom}(\mathbf{Z}, M) \xrightarrow{\mathrm{Hom}(\varepsilon, 1_M)} \mathrm{Hom}(\mathbf{Z}G, M) \xrightarrow{\mathrm{Hom}(i, 1_M)} \mathrm{Hom}(I, M) \to 0$$

$$0 \to \mathrm{Hom}(J, M) \xrightarrow{\mathrm{Hom}(j, 1_M)} \mathrm{Hom}(\mathbf{Z}G, M) \xrightarrow{\mathrm{Hom}(\mu, 1_M)} \mathrm{Hom}(\mathbf{Z}, M) \to 0$$

are exact by 7.5. We observe that $\mathbf{Z} \otimes M \cong M$ and $\mathrm{Hom}(\mathbf{Z}, M) \cong M$ (12.4). Since $\mathbf{Z}G$ is G-free, it is G-induced (13.1). Therefore if A is $\mathbf{Z}G \otimes M$ or $\mathrm{Hom}(\mathbf{Z}G, M)$, then A is G-induced (13.3). Thus any of the exact sequences in (2) is one of the exact sequences

$$0 \to M^- \to A \to M \to 0, \qquad 0 \to M \to A \to M^+ \to 0$$

where M^- is $I \otimes M$ [resp. $\mathrm{Hom}(J, M)$] and M^+ is $J \otimes M$ [resp. $\mathrm{Hom}(I, M)$]. By 15.2, we have the exact sequences

$$0 = H^r(G, A) \to H^r(G, M) \to H^{r+1}(G, M^-) \to H^{r+1}(G, A) = 0$$

and

$$0 = H^{r-1}(G, A) \to H^{r-1}(G, M^+) \to H^r(G, M) \to H^r(G, A) = 0.$$

We conclude that

$$H^{r-1}(G, M^+) \cong H^r(G, M) \qquad \text{and} \qquad H^r(G, M) \cong H^{r+1}(G, M^-).$$

These isomorphisms are called *dimension shifters*. The dimension of the cohomology group is shifted up when the coefficient group M is replaced by $I \otimes M$ or $\mathrm{Hom}(J, M)$, and shifted down when the coefficient group M is replaced by $J \otimes M$ or $\mathrm{Hom}(I, M)$.

17.2 *The uniqueness of $H^0(G, M)$.* In this section we show that $H^0(G, M) = M^G/TM$, and hence is independent of the choice of a complete resolution of G. Let

$\varepsilon(x)x_0) = 0$, i.e.,

$$\varphi(x) = \varepsilon(x)\cdot m' \qquad \text{where} \quad m' = \varphi(x_0) \in M.$$

Conversely any 0-cochain φ defined by $\varphi(x) = \varepsilon(x)\cdot m'$ for some $m' \in M$ is a 0-cocycle. For,

$$(\delta_0\varphi)(x_1) = \varphi(\partial_1 x_1) = \varepsilon\partial_1 x_1 \cdot m' = 0.$$

Since in particular

$$m' = \varepsilon(x_0)m' = \varepsilon(gx_0)m' = g\cdot 1m' = gm',$$

we have $m' \in M^G$. We conclude that the 0-cocycles are the G-homomorphisms $\{\varepsilon(\ )\cdot m\}_{m\in M^G}$. Since $\varepsilon(\ )TM \cong TM$ and $\varepsilon(\ )M^G \cong M^G$, we have shown

$$H^0(G, M) \cong M^G/TM.$$

17.3 *The uniqueness of $H^r(G, M)$.* Let X_* and Y_* be complete resolutions of the finite group G, and $H^r_{X_*}(G, M)$, $H^r_{Y_*}(G, M)$ the corresponding cohomology groups with coefficients in the G-module M. In 17.2 we have shown $H^0_{X_*}(G, M) = H^0_{Y_*}(G, M)$.

CASE $r \geqq 0$ Consider the exact sequence

$$0 \to M \to A \to M^+ \to 0$$

of 17.1 where A is G-induced. Then by dimension shifters $H^r_{X_*}(G, M^+) \cong H^{r+1}_{X_*}(G, M)$ and $H^r_{Y_*}(G, M^+) \cong H^{r+1}_{Y_*}(G, M)$. It follows by induction that $H_{X_*}(G, M) \cong H_{Y_*}(G, M)$ for $r \geqq 0$.

CASE $r \leqq 0$ Consider the exact sequence

$$0 \to M^- \to A \to M \to 0$$

of 17.1. As in case $r \geqq 0$, using dimension shifting and induction we get $H^r_{X_*}(G, M) \cong H^r_{Y_*}(G, M)$ for $r \leqq 0$.

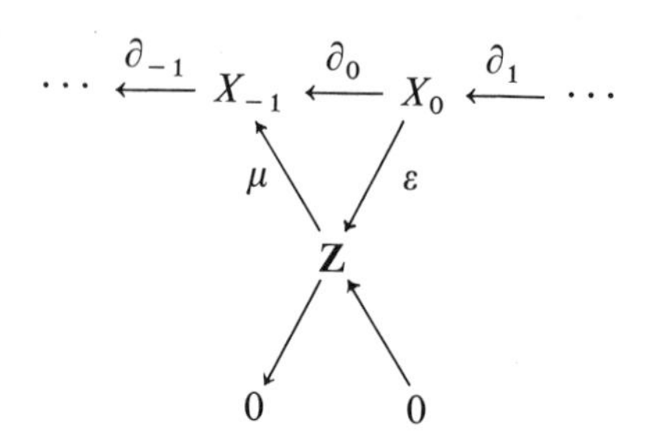

be a complete resolution of G. If $x \in X_0$, $\partial_0 x = \mu\varepsilon(x) = \varepsilon(x)\cdot\mu($
$\varepsilon(x)$ is an integer. Write $\mu(1) = \sum_{\iota=1}^{r} \gamma_\iota y_\iota$ where $\gamma_\iota \in \mathbf{Z}G$ and $\{y_\iota\}$
basis of X_{-1}. We note that $g\gamma_\iota = \gamma_\iota$ since $g\mu(1) = \mu(g1) = \mu(1)$.
if $\gamma_\iota = \sum z_{g,\iota} g$, then the integers $z_{g,\iota}$ are equal for every $g \in G$
$\gamma_\iota = z_\iota T$ for some integer z_ι. Therefore $\mu(1) = T\cdot y$ where

$$y = z_1 y_1 + \cdots + z_r y_r. \tag{3}$$

We contend the integers z_ι are relatively prime. For, let d be their g
common divisor. Then $y = d\cdot y'$ and $0 = \partial_{-1}\mu(1) = d\cdot\partial_{-1}(T\cdot y')$
implies $\partial_{-1}(T\cdot y') = 0$; i.e., $T\cdot y'$ is in the image of μ. Therefore

$$T\cdot y' = \gamma\mu(1) = \gamma\cdot d\cdot T\cdot y'$$

for some $\gamma \in \mathbf{Z}G$. Whence $d = 1$ and $z_1, \ldots, z_r$ are relatively prime
$e_1, \ldots, e_r$ be integers such that

$$e_1 z_1 + e_2 z_2 + \cdots + e_r z_r = 1. \tag{4}$$

Let M be a G-module. For any $m \in M$ define the (-1)-cochain
$\mathrm{Hom}_G(X_{-1}, M)$ by $\varphi_m(y_\iota) = e_\iota m$. Then by (3) and (4) above $\varphi_m(y) =$
Conversely, any (-1)-cochain $\varphi \in \mathrm{Hom}_G(X_{-1}, M)$ corresponds to the
ment $\varphi(y) \in M$. Thus there is a one-to-one correspondence between
(-1)-cochains and the elements of the G-module M. What are the
coboundaries? Let $x \in X_0$ and $\varphi_m \in \mathrm{Hom}_G(X_{-1}, M)$, then

$$(\delta_{-1}\varphi_m)(x) = \varphi_m\partial_0 x = \varphi_m(\varepsilon(x)T\cdot y) = \varepsilon(x)Tm.$$

We conclude that the 0-coboundaries are the G-homomorphisms $\{\varepsilon($
$Tm\}_{m\in M}$. Next, what are the 0-cocycles? Let φ be a 0-cocycle, i.e., $\delta_0\varphi$
$\varphi\ \partial_1 = 0$. Then φ must vanish at every $\partial_1 x_1$, $x_1 \in X_1$. So, for any
cocycle φ, $\varepsilon(x) = 0$ implies $\varphi(x) = 0$. Select $x_0 \in X_0$ such that $\varepsilon(x_0) = 1$
this is possible since ε is onto. We have $\varepsilon(x - \varepsilon(x)x_0) = 0$. Thus $\varphi(x -$

CHAPTER

3

Computations

This chapter is devoted to computations and interpretations of cohomology of groups and comparisons with homology of groups. The concluding sections deal with the special case of cyclic groups.

18. The Computation of $H^{-1}(G, M)$†

Let G be a finite group and M a G-module. By 17.1, we have the dimension-shifting isomorphism $H^r(G, M) = H^{r+1}(G, I \otimes M)$. In particular

$$H^{-1}(G, M) = H^0(G, I \otimes M) = (I \otimes M)^G / T(I \otimes M).$$

In order to obtain another interpretation of $H^{-1}(G, M)$ we first consider $(\mathbf{Z}G \otimes M)^G$. A generator of $\mathbf{Z}G \otimes M$ is of the form $(\sum_{g \in G} z_g g) \otimes m = \sum_{g \in G} g \otimes z_g m$. Moreover

$$\mathbf{Z}G \otimes M = (\sum_{g \in G} \mathbf{Z}_g) \otimes M = \sum_{g \in G} (\mathbf{Z}_g \otimes M) \cong \sum_{g \in G} M_g$$

† See also Section 21.1.

where $\mathbf{Z}_g = \mathbf{Z}$ and $M_g = M$ for every $g \in G$ and the sums are direct. Therefore:

(1) *An element of* $\mathbf{Z}G \otimes M$ *can be represented uniquely in the form* $\sum_{g \in G} g \otimes m_g$ *where* $m_g \in M$.

If $\sum_{g \in G} g \otimes m_g \in (\mathbf{Z}G \otimes M)^G$, then

$$\sum_{g \in G} hg \otimes hm_g = \sum_{g \in G} g \otimes m_g \qquad \text{for every} \quad h \in G.$$

Replacing g by $h^{-1}g$ in the left side of the last identity we get

$$\sum_{g \in G} g \otimes hm_{h^{-1}g} = \sum_{g \in G} g \otimes m_g. \tag{2}$$

By (1) and (2)

$$hm_{h^{-1}g} = m_g. \tag{3}$$

In particular if $h = g$, (3) implies

$$hm_1 = m_h \qquad \text{for every} \quad h \in G. \tag{4}$$

Thus if $\Sigma_{g \in G}\, g \otimes m_g \in (\mathbf{Z}G \otimes M)^G$, then $hm_1 = m_h$ for every $h \in G$. Conversely let $\sum_{g \in G} g \otimes m_g \in \mathbf{Z}G \otimes M$ such that (4) is satisfied. Then

$$\begin{aligned} h \sum_{g \in G} g \otimes m_g &= \sum_{g \in G} hg \otimes hm_g = \sum_{g \in G} hg \otimes hgm_1 \\ &= \sum_{g \in G} g \otimes m_g. \end{aligned}$$

Thus $\sum_{g \in G} g \otimes m_g \in \mathbf{Z}G \otimes M$ satisfies (4) if and only if $\sum_{g \in G} g \otimes m_g \in (\mathbf{Z}G \otimes M)^G$. Recall that $I \otimes M$ is a submodule of $\mathbf{Z}G \otimes M$ by 17.1. So $\sum_{g \in G} g \otimes m_g \in (I \otimes M)^G$ if and only if

$$0 = (\varepsilon \otimes 1_M) \sum_{g \in G} g \otimes m_g = \sum_{g \in G} 1 \otimes m_g = \sum_{g \in G} m_g = \sum_{g \in G} gm_1 = Tm_1$$

where T is the trace map. Let $M_T = \{m \in M \text{ s.t. } Tm = 0\}$, then

$$(I \otimes M)^G \cong M_T. \tag{5}$$

Next we give a similar interpretation of $T(I \otimes M)$. Observe that $I \otimes M$ is $\mathbf{Z}$-generated by the elements $(g - 1) \otimes m$, with $g \in G$, $m \in M$. Moreover

$$T((g - 1) \otimes m) = \sum_{h \in G} hg \otimes hm - \sum_{h \in G} h \otimes hm. \tag{6}$$

Replacing h by hg^{-1} in the first term of the right-hand side we get

$$(7) \qquad T((g-1)\otimes m) = T(1\otimes(g^{-1}m - m)) \in (I\otimes M)^G.$$

Thus an element $T(\sum_{g\in G}(g-1)\otimes m_g)$ of $T(I\otimes M)$ corresponds under the isomorphism (5) to the element

$$\sum_{g\in G} g^{-1}m_g - m_g = \sum_{g\in G}(g^{-1}-1)m_g \in IM.$$

This together with (5) implies

$$H^{-1}(G, M) = M_T/IM.$$

19. The Standard Complete Resolution

The complete resolution

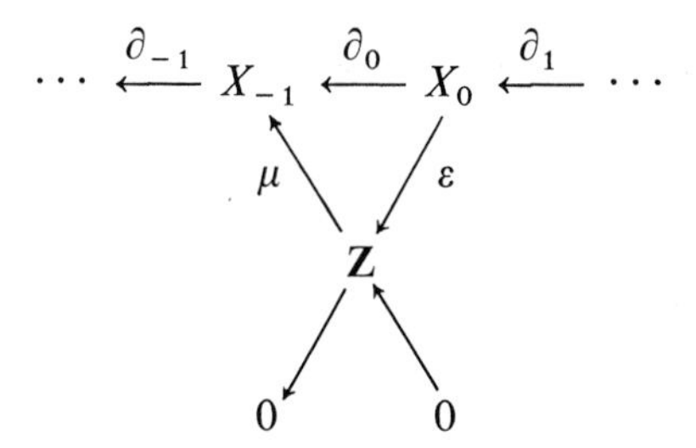

obtained from the bar resolution (16.1) of the G-module $\mathbf{Z}$ by the method given in 16.2 is called the *standard complete resolution of G*. Recall that $X_0 \cong \mathbf{Z}G$ the group ring of G, and X_r for $r = 1, 2, \ldots$, is the free G-module freely generated over $\mathbf{Z}G$ by the symbols $[g_1, \ldots, g_r]$, $g \in G$. Moreover, for $r > 1$,

$$(8) \qquad \partial_r = \sum_{\iota=0}^{r}(-1)^\iota \varepsilon_r^\iota$$

where ε_r^ι are the face maps. Written explicitly

$$\partial_r[g_1, \ldots, g_r] = g_1[g_2, \ldots, g_r] + \sum_{\iota=1}^{r-1}(-1)^\iota[g_1, \ldots, g_\iota g_{\iota+1}, \ldots, g_r] + (-1)^r[g_1, \ldots, g_{r-1}].$$

Recall further that $X_{-r} = \text{Hom}(X_{r-1}, \mathbf{Z})$ and $\partial_{-r} = \text{Hom}(\partial_r, 1_{\mathbf{Z}})$. Finally the augmentation $\varepsilon: \mathbf{Z}G \to \mathbf{Z}$ is given by $\varepsilon[\ ] = 1$, and the coaugmentation $\mu: \mathbf{Z} \to X_{-1}$ by $\mu = \text{Hom}(\varepsilon, 1_{\mathbf{Z}})$ where we have identified $\mathbf{Z}$ with $\text{Hom}(\mathbf{Z}, \mathbf{Z})$. We will now take a closer look at X_{-1} and μ. The G-homomorphism

$$\mu = \text{Hom}(\varepsilon, 1_{\mathbf{Z}}): \text{Hom}(\mathbf{Z}, \mathbf{Z}) \to \text{Hom}(X_0, \mathbf{Z})$$

is completely determined by its value on 1. We have

$$\mu 1 = 1_{\mathbf{Z}} \cdot 1 \cdot \varepsilon = \varepsilon.$$

Next we will express ε in terms of a $\mathbf{Z}$-basis of X_{-1}. Such a basis is the dual basis of the basis $\{g[\ ]\}_{g \in G}$ of X_0. Namely, the $\mathbf{Z}$-homomorphisms $\langle\ \rangle^h: X_0 \to \mathbf{Z}$ defined by

$$\langle\ \rangle^h g[\ ] = h\langle\ \rangle h^{-1} g[\ ] = \begin{cases} 1 & \text{if} \quad h = g \\ 0 & \text{otherwise} \end{cases}$$

form a $\mathbf{Z}$-basis of X_{-1}.

We claim $\varepsilon = \sum_{h \in G} \langle\ \rangle^h$. For, on the $\mathbf{Z}$-basis $\{g[\ ]\}_{g \in G}$ of X_0,

$$\varepsilon(g[\ ]) = 1 \qquad \text{for every} \quad g \in G$$

and

$$\Big(\sum_{h \in G} \langle\ \rangle^h\Big)(g[\ ]) = 1 \qquad \text{for every} \quad g \in G.$$

Thus $\mu(1) = T\langle\ \rangle$. We note further that the single element $\langle\ \rangle$ is a G-basis of X_{-1}.

Similarly for $g, k \in G$, define the $\mathbf{Z}$-homomorphism

$$\langle k \rangle^g: X_1 \to \mathbf{Z}$$

by

$$\langle k \rangle^g(h[l]) = \begin{cases} 1 & \text{if} \quad k = l \text{ and } g = h \\ 0 & \text{otherwise} \end{cases}$$

for every $h, l \in G$. Then $\{\langle k \rangle^g\}_{g,k \in G}$ is a $\mathbf{Z}$-basis of X_{-2}. The G-homomorphism $\partial_{-1}: X_{-1} \to X_{-2}$ is completely determined by its values on the G-basis $\langle\ \rangle$ of X_{-1}. Since $\partial_{-1}\langle\ \rangle = \langle\ \rangle\, \partial_1 \in \text{Hom}(X_1, \mathbf{Z})$, $\partial_{-1}\langle\ \rangle$

can be written in terms of the **Z**-basis $\{\langle k\rangle^g\}_{g,k\in G}$ of X_{-2}. We claim

$$\partial_{-1}\langle\ \rangle = \sum_{g\in G}(\langle g\rangle^{g^{-1}} - \langle g\rangle). \tag{9}$$

In fact

$$(\partial_{-1}\langle\ \rangle)(h[l]) = \langle\ \rangle\,\partial_1(h[l]) = \langle\ \rangle hl[\ ] - \langle\ \rangle h[\ ].$$

Therefore

$$(\partial_{-1}\langle\ \rangle)(h[l]) = \begin{cases} 1 & \text{if } hl = 1 \text{ and } h \neq 1 \\ 0 & \text{if } hl = 1 \text{ and } h = 1 \\ -1 & \text{if } hl \neq 1 \text{ and } h = 1 \\ 0 & \text{if } hl \neq 1 \text{ and } h \neq 1. \end{cases}$$

Whereas

$$\sum_{g\in G}(\langle g\rangle^{g^{-1}} - \langle g\rangle)(h[l]) = \begin{cases} 1 & \text{if } hl = 1 \text{ and } h \neq 1 \\ 0 & \text{if } hl = 1 \text{ and } h = 1 \\ -1 & \text{if } hl \neq 1 \text{ and } h = 1 \\ 0 & \text{if } hl \neq 1 \text{ and } h \neq 1. \end{cases}$$

Finally define the **Z**-homomorphism

$$\langle g_1, \ldots, g_r\rangle^g : X_r \to \mathbf{Z}$$

for $r \geqq 1$ by

$$(\langle g_1, \ldots, g_r\rangle^g)(h[h_1, \ldots, h_r]) = \begin{cases} 1 & \text{if } g_\iota = h_\iota \text{ and } g = h \\ 0 & \text{otherwise.} \end{cases}$$

Then $\{\langle g_1, \ldots, g_r\rangle^g\}_{g_\iota\in G, g\in G}$ is a **Z**-basis of X_{-r-1}. Define the G-homomorphisms $\varepsilon_\iota^r: X_{-r} \to X_{-r-1}$ called *coface maps* by

$$\varepsilon_0^r\langle g_1, \ldots, g_{r-1}\rangle = \sum_{g\in G}\langle g, g_1, \ldots, g_{r-1}\rangle^{g^{-1}}$$

$$\varepsilon_\iota^r\langle g_1, \ldots, g_{r-1}\rangle = \sum_{g\in G}\langle g_1, \ldots, g_\iota g^{-1}, g, g_{\iota+1}, \ldots, g_{r-1}\rangle \quad \text{for } 0 < \iota < r$$

$$\varepsilon_r^r\langle g_1, \ldots, g_{r-1}\rangle = \sum_{g\in G}\langle g_1, \ldots, g_{r-1}, g\rangle.$$

It can be shown by a method similar to the proof of (9) that

$$\partial_{-r} = \sum_{\iota=0}^{r} (-1)^{\iota} \varepsilon_{\iota}^{r}. \tag{10}$$

19.1 *The computation of $H^1(G, M)$.* Using the standard complete resolution of G, a 1-cocycle is a G-homomorphism $\varphi: X_1 \to M$ such that

$$\begin{aligned}\delta^1\varphi[g_1, g_2] &= \varphi\, \partial_1[g_1, g_2] \\ &= g_1\varphi[g_2] - \varphi[g_1g_2] + \varphi[g_1] = 0.\end{aligned}$$

Namely, a 1-cocycle φ restricted to the G-basis $\{[g]\}_{g\in G}$ of X_1 may be viewed as a map

$$\varphi: G \to M$$

from G to the G-module M such that

$$\varphi(g_1g_2) = \varphi(g_1) + g_1\varphi(g_2).$$

Such maps are called *crossed homomorphisms.* A 1-coboundary is a G-homomorphism

$$\delta^0\psi: X_1 \to M$$

where $\psi: X_0 \to M$ is a 0-cochain. Thus a 1-coboundary restricted to the G-basis $\{[g]\}_{g\in G}$ of X_1 may be viewed as a map

$$\delta^0\psi: G \to M$$

defined by $(\delta^0\psi)[g] = \psi\partial_1[g] = g\psi[\;] - \psi[\;]$. Such maps $\pi: G \to M$ defined by $\pi(g) = gm - m$ for some $m \in M$ are called *principal homomorphisms.* Therefore

$$H^1(G, M) \cong \text{Crossed homomorphisms/Principal homomorphisms.}$$

Thus, *in the special case where the action of G on M is trivial we have*

$$H^1(G, M) \cong \mathscr{G}(G, M) \tag{11}$$

where $\mathscr{G}(G, M)$ is the set of homomorphisms from the group G to the abelian group M. Finally we remark that the abelian group structure of $H^1(G, M)$ corresponds under the isomorphism (11) to the abelian structure of $\mathscr{G}(G, M)$ induced by the $\mathbf{Z}$-module structure of M.

19.2 *The normalized standard complete resolution of G.* If in the construction of the bar resolution of the G-module $\mathbf{Z}$ we delete the elements $[g_1, \ldots, g_r]$ from the G-basis of X_r whenever $g_\iota = 1$ for some $\iota = 1, \ldots, r$, and modify the faces and the splitting homotopies as in (i), (ii), and (iii) below, the resulting subsequence of the bar resolution is called the *normalized bar resolution* of the G-module $\mathbf{Z}$. The modifications of the faces and the splitting homotopies are as follows:

(i) The augmentation $\varepsilon\colon X_0 \to \mathbf{Z}$ and the $\mathbf{Z}$-homomorphism $E\colon \mathbf{Z} \to X_0$ are unchanged since X_0 is unchanged, i.e., $\varepsilon[\ ] = 1$ and $E(1) = [\ ]$.

(ii)
$$\varepsilon_r^0[g_1, \ldots, g_r] = g_1[g_2, \ldots, g_r]$$

$$\varepsilon_r^\iota[g_1, \ldots, g_r] = \begin{cases} [g_1, \ldots, g_\iota g_{\iota+1}, \ldots, g_r] & \text{if } g_\iota g_{\iota+1} \neq 1 \\ 0 & \text{if } g_\iota g_{\iota+1} = 1 \end{cases}$$

$$\varepsilon_r^r[g_1, \ldots, g_r] = [g_1, \ldots, g_{r-1}].$$

(iii)
$$D_r(g[g_1, \ldots, g_r]) = \begin{cases} [g, g_1, \ldots, g_r] & \text{if } g \neq 1 \\ 0 & \text{if } g = 1. \end{cases}$$

The verification of the acyclicity of the normalized bar resolution is similar to that of the bar resolution, given in 16.1, with the obvious slight changes in the detail. The complete resolution of G obtained from the normalized bar resolution of the G-module $\mathbf{Z}$ by the method of 16.2 is called the *normalized standard complete resolution* of G.

N.B. In practice when dealing with the normalized case it is accommodative to use the formulas for the standard complete resolution with the understanding that $[g_1, \ldots, g_r]$ is synonymous with zero whenever $g_\iota = 1$ for some $\iota = 1, \ldots, r$. This amounts to the same thing as using the proper formulas (i), (ii), and (iii) above.

20. Extensions

In this section we investigate the relation between $H^2(G, M)$ and extensions of the abelian group M by the group G.

20.1 *Compatible extensions.* Let M be an abelian group and G a group. An exact sequence

$$0 \to M \xrightarrow{i} E \xrightarrow{j} G \to 1 \tag{12}$$

of homomorphisms of groups is called an *extension* of M by G. We will write the compositions in E and G multiplicatively and in M additively. An extension of M by G canonically furnishes M with a structure of G-module. In fact, let $\rho: G \to E$ be a map from the underlying set of G to the underlying set of E such that

$$\rho(1) = 1 \quad \text{and} \quad j\rho = 1_G.$$

Let Aut M denote the group of automorphisms of M. If $e \in E$, then $e(im)e^{-1}$ is not in general im, however $j(e(im)e^{-1}) = 1$. Therefore there exists an element (denote it) $em \in M$ such that

$$i(em) = e(im)e^{-1}.$$

This yields a homomorphism $\theta: E \to \text{Aut } M$. We observe that $M \xrightarrow{i} E \xrightarrow{\theta} \text{Aut } M$ is the trivial map since M is abelian, i.e., $\theta(im) = 1_M$ for every $m \in M$. Hence θ induces a homomorphism

$$\varphi: G \to \text{Aut } M$$

defined by

$$i(\varphi(g)m) = \rho(g)(im)\rho(g)^{-1}. \tag{13}$$

If we write $\varphi(g)m = gm$, then (13) furnishes M with a G-module structure, i.e.,

$$1m = m, \qquad m \in M$$

$$g(m_1 + m_2) = gm_1 + gm_2, \qquad g \in G, \qquad m_1, m_2 \in M,$$

$$(gh)m = g(hm), \qquad g, h \in G, \qquad m \in M.$$

We must show that this action of G on M is independent of the map ρ. Suppose $\rho': G \to E$ is another map such that $\rho'(1) = 1$ and $j\rho' = 1_G$. Then $\rho'(g) = \rho(g)im'$ for some $m' \in M$. Therefore

$$\rho'(g)(im)\rho'(g)^{-1} = \rho(g)(im)\rho(g)^{-1}$$

since M is abelian. Hence, by (13), the action of G on M is independent of the map ρ.

If M is a G-module, we say an extension

$$0 \to M \to E \to G \to 1$$

is *compatible* with the G-module structure of M if the G-module structure obtained from the extension coincides with the given G-module structure of M.

Define a relation among the compatible extensions of M by G as follows:

Given the compatible extensions

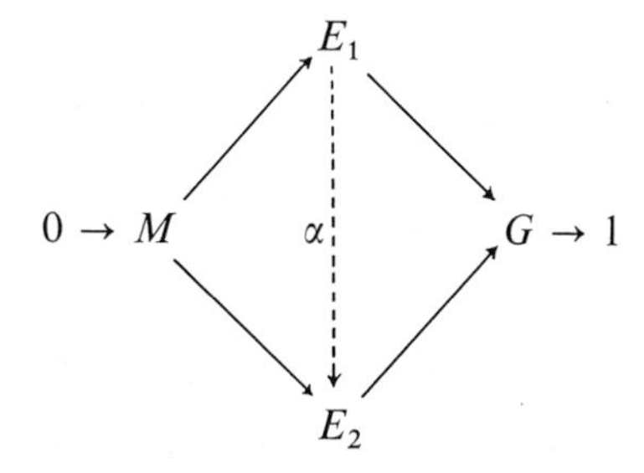

we say the extension above is related to the extension below if there exists a homomorphism $\alpha: E_1 \to E_2$ such that the completed diagram is commutative. One can show by diagram chasing (or the Five Lemma) that α is an isomorphism. Hence this relation is an equivalence relation. We denote by $\mathscr{E}(G, M)$ the equivalence classes of compatible extensions of M by G modulo this relation.

20.2 *The computation of $H^2(G, M)$.* In this section X_* is the normalized standard complete resolution of G, and M is a G-module. Let

$$M \times G = \{(m, g): m \in M, g \in G\}.$$

Let φ be a normalized 2-cocycle, i.e., $\varphi: X_2 \to M$ is a G-homomorphism such that

$$(14) \qquad (\delta^2\varphi)[g_1, g_2, g_3] = g_1\varphi[g_2, g_3] - \varphi[g_1g_2, g_3] + \varphi[g_1, g_2g_3) - \varphi[g_1, g_2] = 0.$$

(It is understood that in (14) the symbol $[g, h] = 0$ whenever $g = 1$ or $h = 1$.) A normalized 2-cocycle is called a *factor set*. We define a multiplicative structure on the set $M \times G$ by

$$(m, g)(n, h) = (m + gn + \varphi[g, h], gh) \tag{15}$$

for every $(m, g), (n, h) \in M \times G$. We shall denote $M \times G$ with this multiplicative structure by $M \times_\varphi G$. We claim $M \times_\varphi G$ is a group. For,

(i) $(0, 1)$ *is a left identity:*
$(0, 1)(n, h) = (n + \varphi[1, h], h) = (n, h)$ since φ is normalized.
$(0, 1)$ *is a right identity:*
$(m, g)(0, 1) = (m + \varphi[g, 1], g) = (m, g)$ since φ is normalized.
(ii) *Associativity:*

$$(m_1, g_1)(m_2, g_2))(m_3, g_3) = (m_1 + g_1m_2 + \varphi[g_1, g_2] + g_1g_2m_3 + \varphi[g_1g_2, g_3], g_1g_2g_3) \tag{16}$$

and

$$(m_1, g_1)((m_2, g_2)(m_3, g_3)) = (m_1 + g_1m_2 + g_1g_2m_3 + g_1\varphi[g_2, g_3] + \varphi(g_1, g_2g_3), g_1g_2g_3). \tag{17}$$

The right sides of (16) and (17) are identical by (14).
(iii) *Inverses:*
$(-g^{-1}m - \varphi[g^{-1}, g], g^{-1})$ is the left inverse of (m, g). For this to be the right inverse also, we must have

$$g\varphi[g^{-1}, g] = \varphi]g, g^{-1}].$$

This identity follows from (14) when g_1, g_2, g_3 are replaced by g, g^{-1}, g, respectively.

Define the group homomorphisms i, j,

$$i: M \to M \times_\varphi G$$

by $im = (m, 1)$, and

$$j: M \times_\varphi G \to G$$

by $j(m, g) = g$. Then the sequence

$$0 \to M \xrightarrow{i} M \times_\varphi G \xrightarrow{j} G \to 1 \tag{18}$$

is exact. Moreover (18) is a compatible extension of the G-module M. For, let $\rho: G \to M \times_{\varphi} G$ be defined by $\rho g = (0, g)$. Then $\rho(1) = (0, 1)$, $j\rho = 1_G$, and

$$\begin{aligned}(0, g)(m, 1)(0, g)^{-1} &= (0, g)(m, 1)(-\varphi[g^{-1}, g], g^{-1}) \\ &= (gm, 1) \qquad \text{since} \quad \delta\varphi[g, g^{-1}, g] = g\varphi[g^{-1}, g] - \varphi[g, g^{-1}] = 0 \\ &= i(gm).\end{aligned}$$

Summarizing, we have constructed a map Φ from the normalized 2-cocycles to $\mathscr{E}(G, M)$ which assigns to the 2-cocycle φ the element of $\mathscr{E}(G, M)$ represented by the extension $0 \to M \to M \times_{\varphi} G \to G \to 1$. We claim Φ induces a map

$$\overline{\Phi}: H^2(G, M) \to \mathscr{E}(G, M).$$

In fact, Φ assigns to the normalized 2-cocycle $\varphi + \delta^1\psi$ the extension

$$0 \to M \to M \times_{\varphi + \delta^1\psi} G \to G \to 1. \tag{19}$$

The extensions (18) and (19) belong to the same equivalence class. For, if the map

$$\alpha: M \times_{\varphi} G \to M \times_{\varphi + \delta^1\psi} G$$

is defined by $\alpha(m, g) = (m - \psi[g], g)$, then α is a homomorphism and the diagram

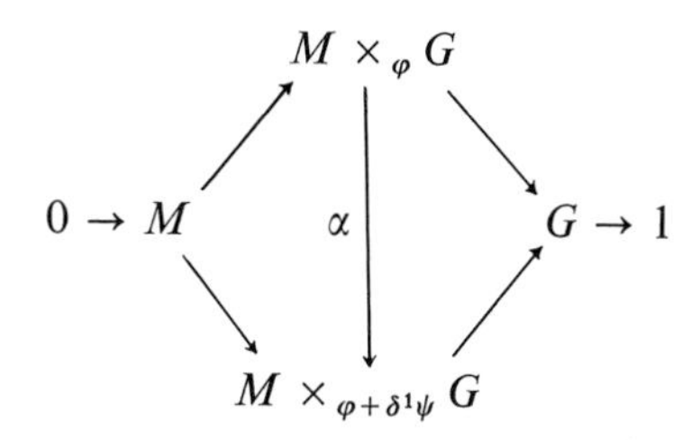

is commutative since ψ is a normalized 1-cochain. Hence $\overline{\Phi}$ is well defined. Next we claim $\overline{\Phi}$ is a monomorphism. Suppose the normalized 2-cocycles φ and φ' are mapped by Φ to equivalent extensions, i.e., the diagram

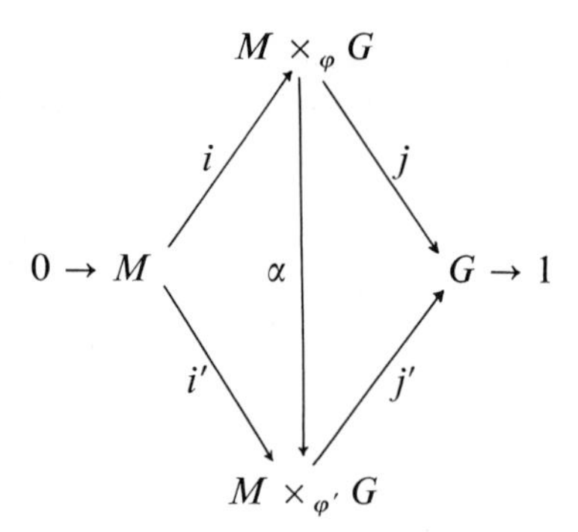

is commutative. In what follows we will distinguish the elements of $M \times_{\varphi} G$ from the elements of $M \times_{\varphi'} G$ by the subscripts φ and φ', respectively, e.g., $(m, g)_{\varphi} \in M \times_{\varphi} G$. Since $j = j'\alpha$ and $\alpha i = i'$, we have $\alpha(0, g)_{\varphi} = (\psi[g], g)_{\varphi'}$ where $\psi[1] = 0$, i.e., ψ is a normalized 1-cochain

$$\begin{aligned}\alpha((0, g)_{\varphi} \cdot (0, h)_{\varphi}) &= (\psi[g], g)_{\varphi'} \cdot (\psi[h], h)_{\varphi'} \\ &= (\psi[g] + g\psi[h] + \varphi'[g, h], gh)_{\varphi'}.\end{aligned}$$

But

$$\begin{aligned}\alpha((0, g)_{\varphi} \cdot (0, h)_{\varphi}) &= \alpha((\varphi[g, h], gh)_{\varphi}) \\ &= (\varphi[g, h] + \psi[gh], gh)_{\varphi'}.\end{aligned}$$

Hence

$$\begin{aligned}\varphi[g, h] &= g\psi[h] - \psi[gh] + \psi[g] + \varphi'[g, h] \\ &= (\delta^1\psi + \varphi')[g, h].\end{aligned}$$

Finally we claim that $\overline{\Phi}$ is an epimorphism. Let

$$0 \to M \xrightarrow{i} E \underset{\rho}{\overset{j}{\rightleftarrows}} G \to 1$$

be an extension compatible with the action of G on M. Construct the map $\rho: G \to E$ such that $\rho[1] = 1$ and $j\rho = 1_G$. Then $\rho(g)\rho(h) \cdot (\rho(gh)^{-1}) \in iM$. Let $\varphi[g, h] \in M$ such that

$$\rho(g)\rho(h) = i\varphi[g, h] \cdot \rho(gh). \tag{20}$$

φ is a map, $\varphi: G \times G \to M$, and hence is a 2-cochain (it is a map from the G-generators of X_2 to M); φ is a normalized 2-cochain since $\rho(1) = 1$ and i is a monomorphism; φ is a 2-cocycle, for, every element of E can be written uniquely in the form $im \cdot \rho(g)$. Moreover,

$$\begin{aligned}
&((im_1 \cdot \rho g_1)(im_2 \cdot \rho g_2))(im_3 \cdot \rho g_3) \\
&\quad = (im_1 \cdot i(g_1 m_2) \cdot \rho g_1 \cdot \rho g_2)(im_3 \cdot \rho g_3) \qquad \text{by (13)} \\
&\quad = (i(m_1 + g_1 m_2 + \varphi[g_1, g_2]) \cdot \rho(g_1 g_2))(im_3 \cdot \rho g_3) \qquad \text{by (20)} \\
&\quad = i(m_1 + g_1 m_2 + \varphi[g_1, g_2] + g_1 g_2 \cdot m_3 + \varphi[g_1 g_2, g_3]) \cdot \rho(g_1 g_2 g_3) \\
&\qquad \text{by (20).}
\end{aligned}$$

Similarly

$$\begin{aligned}
&(im_1 \cdot \rho g_1)((im_2 \cdot \rho g_2)(im_3 \cdot \rho g_3)) \\
&\quad = i(m_1 + g_1 m_2 + g_1 g_2 m_3 + g_1 \varphi[g_2, g_3] + \varphi[g_1, g_2 g_3]) \cdot \rho(g_1 g_2 g_3).
\end{aligned}$$

Whence by the associativity of the composition in E we have

$$g_1 \varphi[g_2, g_3] + \varphi[g_1, g_2 g_3] = \varphi[g_1, g_2] + \varphi[g_1 g_2, g_3],$$

i.e., φ is a 2-cocycle. Summarizing, we have shown that:

There exists an isomorphism

$$\overline{\Phi}: H^2(G, M) \to \mathscr{E}(G, M)$$

where $\mathscr{E}(G, M)$ is the equivalence classes of those extensions of M by G which are compatible with the given action of G on M.

In conclusion we remark that the isomorphism $\overline{\Phi}$ endows $\mathscr{E}(G, M)$ with the structure of the abelian group $H^2(G, M)$. The zero element of $\mathscr{E}(G, M)$ corresponds to the zero factor set. Therefore it is the extension

$$0 \to M \to M \times_0 G \to G \to 1.$$

The composition in $M \times_0 G$ thus is given by

$$(m, g)(n, h) = (m + gn, gh).$$

$M \times_0 G$ is called the *semidirect* product of M and G.

21. The Negative Cochain Complex

Let G be a finite group and M a G-module. Let $M^r = \mathrm{Hom}_G(X_r, M)$ where X_* is the standard complete resolution of G. Let $\delta^r = \mathrm{Hom}_G(\partial_{r+1}, 1_M)$. Then, by 14.3,

$$\cdots \longrightarrow M^{r-1} \xrightarrow{\delta^{r-1}} M_r \xrightarrow{\delta^r} M^{r+1} \longrightarrow \cdots$$

is a cochain complex whose cohomology groups are the groups $H^r(G, M)$. We will examine the negative part of this cochain complex in the hope of computing some of the negative dimensional cohomology groups.

For $m \in M$ and $r \geqq 0$, let $m * g[g_1, \ldots, g_r]: X_{-r-1} \to M$ be defined by

$$(21) \qquad (m * g[g_1, \ldots, g_r])\langle h_1, \ldots, h_r\rangle^k = \sum_{h\in G} \langle h_1, \ldots, h_r\rangle^k hg[g_1, \ldots, g_r]\cdot hm.$$

$m * g[g_1, \ldots, g_r]$ is clearly a **Z**-homomorphism. To show it is a G-homomorphism let $k \in G$, then

$$k\langle h_1, \ldots, h_r\rangle k^{-1} = \langle h_1, \ldots, h_r\rangle k^{-1}$$

since **Z** is a trivial G-module. Therefore

$$\begin{aligned} m * g[g_1, \ldots, g_r]\langle h_1, \ldots, h_r\rangle^k &= \sum_{h\in G} \langle h_1, \ldots, h_r\rangle k^{-1}hg[g_1, \ldots, g_r]hm \\ &= \sum_{h\in G} \langle h_1, \ldots, h_r\rangle hg[g_1, \ldots, g_r]khm \\ &= k(m * g[g_1, \ldots, g_r])\langle h_1, \ldots, h_r\rangle \end{aligned}$$

since $\langle h_1, \ldots, h_r\rangle h[g_1, \ldots, g_r]$ is an integer. It follows that

(i) $\quad m * g[g_1, \ldots, g_r]$ is a $(-r-1)$-cochain.

(ii) $\quad (m_1 + m_2)*g[g_1, \ldots, g_r] = m_1*g[g_1, \ldots, g_r] + m_2 * g[g_1, \ldots, g_r].$

(iii) $\quad m*g[g_1, \ldots, g_r] = g^{-1}m * [g_1, \ldots, g_r].$

The verification of the last identity is as follows:

$$\begin{aligned}(m * g[g_1, \ldots, g_r]\langle h_1, \ldots, h_r\rangle &= \sum_{h\in G} \langle h_1, \ldots, h_r\rangle hg[g_1, \ldots, g_r]hm \\ &= \sum_{h\in G} \langle h_1, \ldots, h_r\rangle h[g_1, \ldots, g_r]h(g^{-1}m) \\ &= (g^{-1}m * [g_1, \ldots, g_r])\langle h_1, \ldots, h_r\rangle.\end{aligned}$$

Finally we extend formula (21) to

$$\text{(iv)} \qquad m * \sum_{\iota=1}^{n} h_\iota[g_1, \ldots, g_r] = \sum_{\iota=1}^{n} m * h_\iota[g_1, \ldots, g_r].$$

We will now examine the elements of $M^{-r-1} = \mathrm{Hom}_G(X_{-r-1}, M)$. Recall that $\{[g_1, \ldots, g_r]\}_{g_\iota \in G}$ is a G-basis of X_r and $\{\langle g_1, \ldots, g_r\rangle\}_{g_\iota\in G}$ is the dual G-basis of X_{-r-1}; i.e.,

$$\langle g_1, \ldots, g_r\rangle h[h_1, \ldots, h_r] = \begin{cases} 1 & \text{if } \; g_\iota = h_\iota \text{ and } h = 1 \\ 0 & \text{otherwise.}\end{cases}$$

Therefore

$$(m * [g_1, \ldots, g_r])\langle h_1, \ldots, h_r\rangle = \begin{cases} m & \text{if } \; g_\iota = h_\iota \\ 0 & \text{otherwise.}\end{cases}$$

So, the G-homomorphism

$$\sum_{g_{\iota_j}\in G} m_\iota * [g_{\iota_1}, \ldots, g_{\iota_r}] \qquad \text{where} \quad \iota = (\iota_1, \ldots, \iota_r)$$

sends $\langle g_{\kappa_1}, \ldots, g_{\kappa_r}\rangle$ to m_κ and all other G-basis elements of X_{-r-1} to zero. We conclude that:

Every element of $\mathrm{Hom}_G(X_{-r-1}, M)$ *can be written uniquely in the form*

$$\sum_{g_{\iota_j}\in G} m_\iota * [g_{\iota_1}, \ldots, g_{\iota_r}].$$

Next we express the negative coboundaries $\delta^{-r-1}\colon M^{-r-1} \to M^{-r}$ in terms of the positive boundaries $\partial_r\colon X_r \to X_{r-1}$. For $r > 0$.

$$\varepsilon^{*,\iota}_{-r-1}\colon \mathrm{Hom}_G(X_{-r-1}, M) \to \mathrm{Hom}_G(X_{-r}, M)$$

is defined by $\varepsilon^{*,\iota}_{-r-1} = \mathrm{Hom}_G(\varepsilon^r_\iota, 1_M)$ where ε^r_ι is the ith coface map defined in Section 19. Thus

$$\begin{aligned}(\varepsilon^{*,\iota}_{-r-1}(m * [g_1, \ldots, g_r]))\langle h_1, \ldots, h_{r-1}\rangle \\ &= (m * [g_1, \ldots, g_r])\varepsilon^r_\iota\langle h_1, \ldots, h_{r-1}\rangle \\ &= \sum_{k\in G} \varepsilon^r_\iota\langle h_1, \ldots, h_{r-1}\rangle k[g_1, \ldots, g_r]km \\ &= \sum_{k\in G} \langle h_1, \ldots, h_{r-1}\rangle \varepsilon^\iota_r k[g_1, \ldots, g_r]km\end{aligned}$$

where ε^ι_r is the rth face defined in 16.1,

$$= (m * \varepsilon^\iota_r[g_1, \ldots, g_r])\langle h_1, \ldots, h_{r-1}\rangle.$$

Since for $r = 1, 2, \ldots,$

$$\delta^{-r-1} = \sum_{\iota=0}^{r} (-1)^\iota \varepsilon^{*,\iota}_{-r-1},$$

we have shown

$$\delta^{-r-1}(m * [g_1, \ldots, g_r]) = m * \partial_r[g_1, \ldots, g_r]. \tag{22}$$

In the case $r = 0$ we have

$$\delta^{-1}(m * [\ ]) = \mathrm{Hom}_G(\partial_0, 1_M) = m * [\ ]\partial_0.$$

So

$$\begin{aligned}\delta^{-1}(m * [\ ])[\ ] &= (m * [\ ])\partial_0[\ ] \\ &= m * [\ ]\langle\ \rangle^T \\ &= \sum_{g\in G} \langle\ \rangle^T g[\ ]gm = Tm.\end{aligned} \tag{23}$$

21.1 *Another computation of $H^{-1}(G, M)$.* We have shown in Section 18 that for a finite group G, $H^{-1}(G, M) = M_T/IM$. We give a short proof of this isomorphism with the aid of the formulas above. By (23) the (-1)-cocycles are all the elements $m * [\ ]$ of $\mathrm{Hom}_G(X_{-1}, M)$ such that $\delta^{-1}(m * [\ ])[\ ] = Tm = 0$. Thus the group of (-1)-cocycles is isomorphic to $M_T = \{m \in M \text{ s.t. } Tm = 0\}$. The (-1)-coboundaries are the elements

$$\delta^{-2} \sum_{g\in G} m_g * [g] \in \mathrm{Hom}_G(X_{-1}, M).$$

Since

$$\delta^{-2} \sum_{g \in G} m_g * [g] = \sum m_g * \partial_1[g]$$
$$= \sum_{g \in G} (g^{-1}m_g - m_g) * [\ ],$$

the group of (-1)-coboundaries is isomorphic to IM where I is the kernel of the augmentation map ε. We conclude that $H^{-1}(G, M) = M_T/IM$.

21.2 *The computation of $H^{-2}(G, M)$ (trivial action).* We are unable to give a reasonable interpretation of $H^{-2}(G, M)$ when the action of G on M is arbitrary. However, when this action is trivial we show $H^{-2}(G, M) = M \otimes G/[G, G]$, where $[G, G]$ is the commutator subgroup of G. In what follows we use the normalized standard complete resolution of G.

Every (-2)-cochain $\sum_{g \in G} m_g * [g]$ is a cocycle since its coboundary, $\sum_{g \in G} (g^{-1}m_g - m_g) * [\ ]$ is zero by the triviality of the action. A (-2)-coboundary is a sum of elements of the form

$$\delta^{-3}(m * [g, h]) = m * [h] - m * [gh] + m * [g].$$

Since

$$m * [gh] = m * [g] + m * [h] + (-m * [h] + m * [gh] - m * [g]),$$

we have

(24) *$m * [gh]$ is congruent to $m * [g] + m * [h]$ modulo coboundaries*

and

(25) *$m * [ghg^{-1}h^{-1}]$ is congruent to*

$$m * [g] + m * [h] - m * [g] - m * [h] = 0 \text{ modulo coboundaries}$$

since we are using the normalized standard complete resolution of G. Define the map

$$\theta \colon M \otimes G/[G, G] \to H^{-2}(G, M)$$

by $\theta(m, g[G, G]) = m * [g] +$ coboundaries. The map θ is well defined by (25); θ is linear in $g[G, G]$ by (24) and linear in m by 21,(ii). We contend $H^{-2}(G, M)$ represents $M \otimes G/[G, G]$. In fact the image of θ generates

$H^{-2}(G, M)$. Moreover given a bilinear map

$$\varphi: M \times G/[G, G] \to N$$

where N is a G-module, there exists a $\mathbf{Z}$-homomorphism $\bar{\varphi}: H^{-2}(G, M) \to N$ defined by

$$\bar{\varphi}\Big(\sum_{g\in G} (m_g * [g] + \text{Coboundaries})\Big) = \sum_{g\in G} \varphi(m_g, g[G, G]).$$

$\bar{\varphi}$ is well defined for

$$\begin{aligned}\bar{\varphi}(m * [h] - m * [gh] + m * [g] + \text{Coboundaries}) \\ &= \varphi(m, h[G, G]) - \varphi(m, gh[G, G]) + \varphi(m, g[G, G]) \\ &= \varphi(m, h(gh)^{-1}g[G, G]) \\ &= 0 \qquad \text{since} \quad \varphi \text{ is bilinear.}\end{aligned}$$

Furthermore the diagram

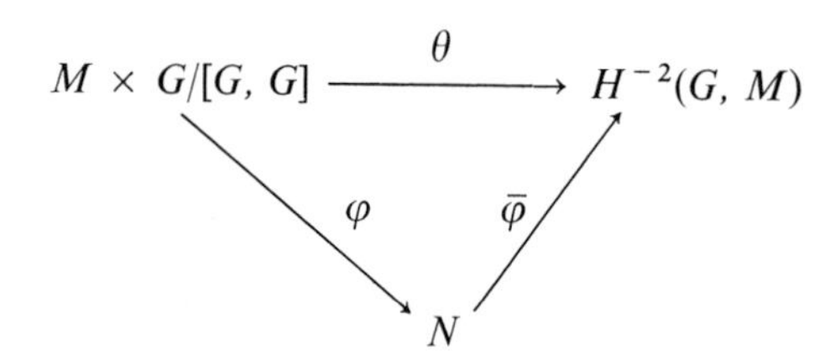

is commutative; i.e.,

$$\bar{\varphi}\theta(m, g[G, G]) = \bar{\varphi}(m * [g] + \text{Coboundaries}) = \varphi(m, g[G, G]).$$

Therefore

$$H^{-2}(G, M) = M \otimes G/[G, G].$$

In particular when M is the trivial G-module $\mathbf{Z}$ we have

$$H^{-2}(G, \mathbf{Z}) = G/[G, G]. \tag{26}$$

21.3 *Direct computation of* $H^{-2}(G, \mathbf{Z})$. The isomorphism (26) can be established directly from $H^{-1}(G, I)$ and dimension shifting where I is the kernel of the augmentation map $\varepsilon: \mathbf{Z}G \to \mathbf{Z}$.

The exact sequence

$$0 \to I \to \mathbf{Z}G \to \mathbf{Z} \to 0$$

of 17.1 gives the isomorphism

$$H^{-2}(G, \mathbf{Z}) \cong H^{-1}(G, I)$$

since $\mathbf{Z}G$ is G-induced. By Section 18 or 21.1 we have

$$H^{-2}(G, \mathbf{Z}) \cong I_T/I^2.$$

We claim $I_T = I$. For, $\sum z_g g \in I$ if and only if $\sum z_g = 0$. Hence for any $\sum z_g g \in I$, $T(\sum z_g g) = (\sum z_g)T = 0$. Thus

$$H^{-2}(G, \mathbf{Z}) \cong I/I^2.$$

The isomorphism (26) now follows from the isomorphism of groups

$$I/I^2 \cong G/[G, G]. \tag{27}$$

To establish this isomorphism, we observe first that I is an abelian group written additively. Define the map

$$\Phi: G \to I/I^2$$

by $\Phi(g) = (g - 1) + I^2$; Φ is a homomorphism since

$$\begin{aligned}\Phi(gh) &= (gh - 1) + I^2 \\ &= (g - 1) + (h - 1) + (g - 1)(h - 1) + I^2 \\ &= (g - 1) + (h - 1) + I^2.\end{aligned}$$

Since I/I^2 is abelian, Φ induces the homomorphism $\overline{\Phi}: G/[G, G] \to I/I^2$ given by $\overline{\Phi}(g[G, G]) = (g - 1) + I^2$. Next define the map

$$\Psi: I \to G/[G, G]$$

by $\Psi(g - 1) = g[G, G]$; Ψ is a homomorphism because the elements $g - 1$ form a $\mathbf{Z}$-basis of I. The homomorphism Ψ induces a homomorphism $\overline{\Psi}: I/I^2 \to G/[G, G]$, for,

$$\begin{aligned}\Psi((g - 1)(h - 1)) &= \Psi((gh - 1) - (g - 1) - (h - 1)) \\ &= gh[G, G]\cdot g^{-1}[G, G]\cdot h^{-1}[G, G] \\ &= ghg^{-1}h^{-1}[G, G] \\ &= [G, G].\end{aligned}$$

It now follows from the definitions that $\overline{\Phi}\overline{\Psi} = 1_{I/I^2}$ and $\overline{\Psi}\overline{\Phi} = 1_{G/[G,G]}$.

22. Homology and Cohomology of Monoids

In this section we define homology and cohomology of monoids using a particular resolution of $\mathbf{Z}$. The proof that homology and cohomology are independent of this resolution follows from the "comparison theorem" in homological algebra (Cartan-Eilenberg [15, Chapter 5, Proposition 1.2]). Essentially one must show that any two $\mathbf{Z}G$-projective resolutions of $\mathbf{Z}$ are homotopic. This is done by arguments similar to those used in Section 9.

A set M with a composition (written multiplicatively) is called a *monoid* if this composition is associative and M has an element which is both a right and a left identity. Thus if M has the further property that every element has an inverse, then M is a group. Let M be a monoid. We define the monoid ring $\mathbf{Z}M$ as the set of all the formal sums $\sum_{\mu \in M} z_\mu \mu$ where $z_\mu \in \mathbf{Z}$ are zero except for a finite number of them, together with the following composition of such formal sums:

$$\sum_{\mu \in M} z_\mu \mu + \sum_{\mu \in M} z'_\mu \mu = \sum_{\mu \in M} (z_\mu + z'_\mu)\mu$$

and

$$\Big(\sum_{\mu \in M} z_\mu \mu\Big)\Big(\sum_{\nu \in M} z'_\nu \nu\Big) = \sum_{\xi \in M} z''_\xi \xi$$

where z''_ξ is the sum of those products $z_\mu z'_\nu$ for which $\mu\nu = \xi$. We say the abelian group A has the structure of a left $\mathbf{Z}M$-module if an action of M on A is given, i.e., a map $\theta\colon M \times A \to A$ with $\theta(\mu, a) = \mu a$ is given such that

$$1a = a, \qquad \mu(a_1 + a_2) = \mu a_1 + \mu a_2, \qquad (\mu\nu)a = \mu(\nu a),$$

for all $\mu, \upsilon \in M$, and $a, a_1, a_2 \in A$. In this case we will say A is a left M-module. The definition of a right M-module is similar. View $\mathbf{Z}$ as an M-module with trivial action, and construct the normalized bar resolution of the M-module $\mathbf{Z}$ exactly as we constructed the normalized bar resolution of the G-module $\mathbf{Z}$ in 16.1. Let

$$X_*\colon 0 \leftarrow \mathbf{Z} \stackrel{\varepsilon}{\leftarrow} X_0 \stackrel{\partial_1}{\leftarrow} X_1 \leftarrow \cdots$$

be the resulting resolution of $\mathbf{Z}$. For any left M-module A the rth cohomology group $H^r(\mathrm{Hom}_{\mathbf{Z}M}(X_*, A))$, $r = 0, 1, \ldots$, of the cochain complex

$$\mathrm{Hom}_{\mathbf{Z}M}(X_0, A) \xrightarrow{\mathrm{Hom}_{\mathbf{Z}M}(\partial_1, 1_A)} \mathrm{Hom}_{\mathbf{Z}M}(X_1, A) \xrightarrow{\mathrm{Hom}_{\mathbf{Z}M}(\partial_2, 1_A)} \cdots$$

is called *the* rth *cohomology group of the monoid M with coefficients in the left M-module A*. We will denote these cohomology groups by $H^r_{\mathcal{M}}(M, A)$.

Thus a 0-cochain is a $\mathbf{Z}M$-homomorphism $\varphi: \mathbf{Z}M \to A$. Hence the 0-cochains can be identified with A. The elements of $H^0_{\mathcal{M}}(M, A)$ are the 0-cocycles, i.e., those $\mathbf{Z}M$-homomorphisms $\varphi: \mathbf{Z}M \to A$ such that $\varphi\partial_1[\mu] = \mu\varphi[\ ] - \varphi[\ ] = 0$. Therefore using the identification of $\mathrm{Hom}_{\mathbf{Z}M}(\mathbf{Z}M, A)$ with A,

$$H^0_{\mathcal{M}}(M, A) = A^M \tag{28}$$

where $A^M = \{a \in A \text{ s.t. } \mu a = a \text{ for every } \mu \in M\}$. For $r = 1, 2, \ldots$, a normalized r-cochain is a $\mathbf{Z}M$-homomorphism $\varphi: X_r \to A$. Since the $[\mu_1, \ldots, \mu_r]$ with $\mu_i \neq 1$ form a basis of X_r, a normalized r-cochain is equivalent to a function $\varphi: M^r \to A$ such that $\varphi(\mu_1, \ldots, \mu_r) = 0$ whenever $\mu_i = 1$ for some $i = 1, \ldots, r$.

In the special case where the monoid M is a group, these cohomology groups are denoted by $H^r(M, A)$ in the literature. However, to avoid confusion we will use the notation $H^r_{\mathcal{M}}(M, A)$ in this section. These cohomology groups were first studied by Eilenberg and MacLane [24].

In the special case where the monoid M is a finite group, it is evident that

$$H^r_{\mathcal{M}}(M, A) \cong H^r(M, A) \qquad \text{for} \quad r = 1, 2, \ldots$$

and for $r = 0$, there exists an epimorphism

$$H^0_{\mathcal{M}}(M, A) \to H^0(M, A),$$

for, $H^0_{\mathcal{M}}(M, A) \cong A^M$ and $H^0(M, A) \cong A^M/TA$. We will return to this comparison in the next section.

To define homology of monoids (or groups) we must first extend the definition of tensor product. Let R be a ring (not necessarily commutative) with an identity element. Let X be a right R-module, and Y a left R-module. Let A be an abelian group. A map $f: X \times Y \to A$ is called *balanced* if

$$f(x_1 + x_2, y) = f(x_1, y) + f(x_2, y)$$

$$f(x, y_1 + y_2) = f(x, y_1) + f(x, y_2)$$

$$f(x, ry) = f(xr, y)$$

for all $r \in R$, x, $x_i \in X$, and y, $y_i \in Y$. We say the abelian group T represents the *tensor product* of X and Y over the ring R if there exists a balanced map $\theta: X \times Y \to T$ such that T is generated by the image of θ, and for any abelian group A and balanced map $f: X \times Y \to A$ there exists a homomorphism of abelian groups $\bar{f}: T \to A$ such that the diagram

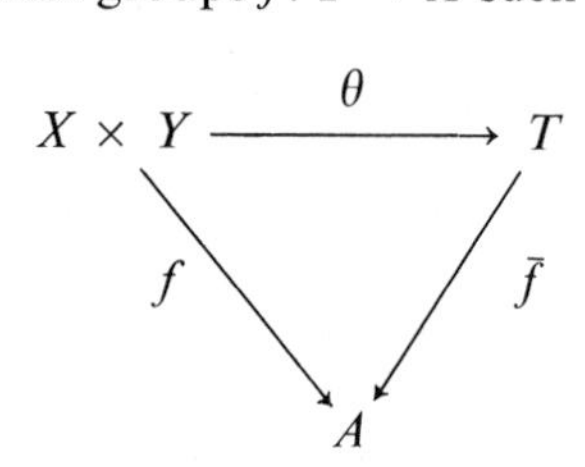

is commutative. Thus the tensor product is unique up to a canonical isomorphism and we shall denote it by $X \otimes_R Y$. The results of Sections 2, 3, and 4 in Chapter 1 can easily be generalized to this case.

Let R be the monoid ring $\mathbf{Z}M$. Let A be a right M-module. Let X_* be the normalized bar resolution of the trivial M-module $\mathbf{Z}$. The rth homology group $H_r(A \otimes_{\mathbf{Z}M} X_*)$, $r = 0, 1, \ldots,$ of the chain complex

$$0 \leftarrow A \otimes_{\mathbf{Z}M} X_0 \xleftarrow{1_A \otimes_{\mathbf{Z}M} \partial_1} A \otimes_{\mathbf{Z}M} X_1 \leftarrow \cdots$$

is called *the* rth *homology group of the monoid* M *with coefficients in the right* M*-module* A. We will denote these homology groups by $H_r^{\mathcal{M}}(M, A)$ or simply $H_r(M, A)$.

22.1 *Comparison of (Tate) cohomology with homology and cohomology of monoids.* Let G be a group and A a left G-module. We can convert A into a right G-module by defining

$$ag = g^{-1}a$$

for all $a \in A$ and $g \in G$. Therefore both $H^r_{\mathcal{M}}(G, A)$ and $H_r^{\mathcal{M}}(G, A)$ are defined for a left G-module A. Now let G be a finite group. Let

$$0 \leftarrow \mathbf{Z} \overset{\varepsilon}{\leftarrow} X_0 \overset{\partial_1}{\leftarrow} X_1 \leftarrow \cdots$$

be the normalized bar resolution of the trivial G-module $\mathbf{Z}$. For the left G-module A consider the diagram

$$(29)\qquad \cdots \to A \otimes_{\mathbf{Z}G} X_1 \xrightarrow{1_A \otimes_{\mathbf{Z}G} \partial_1} A \otimes_{\mathbf{Z}G} X_0 \xrightarrow{\delta'^{-1}} \mathrm{Hom}_{\mathbf{Z}G}(X_0, A) \xrightarrow{\mathrm{Hom}_{\mathbf{Z}G}(\partial_1, 1_A)} \mathrm{Hom}_{\mathbf{Z}G}(X_1, A) \to \cdots$$

where $\delta'^{-1}(a \otimes [\ ]) = Tf_a$ and f_a is the $\mathbf{Z}G$-homomorphism defined by $f_a[\ ] = a$. We note that in the left half of this sequence A stands for the converted right G-module defined above. We relabel this sequence as follows: Let

$$A^r = \mathrm{Hom}_{\mathbf{Z}G}(X_r, A), \qquad \delta'^r = \mathrm{Hom}_{\mathbf{Z}G}(\partial_r, 1_A) \qquad \text{for} \quad r = 0, 1, \ldots$$

$$A^{-r} = A \otimes_{\mathbf{Z}G} X_{r-1} \qquad \text{for} \quad r = 1, 2, \ldots$$

$$\delta'^{-r} = 1_A \otimes_{\mathbf{Z}G} \partial_{r-1} \qquad \text{for} \quad r = 2, 3, \ldots.$$

The sequence (29) is a complex since $\delta'^{r+1}\delta'^r = 0$ for every integer r. We claim there is an isomorphism of cochain complexes

$$\begin{array}{ccccccccc}
\cdots \to & A \otimes_{\mathbf{Z}G} X_1 & \xrightarrow{1_A \otimes_{\mathbf{Z}G} \partial_1} & A \otimes_{\mathbf{Z}G} X_0 & \xrightarrow{\delta'^{-1}} & \mathrm{Hom}_{\mathbf{Z}G}(X_0, A) & \xrightarrow{\mathrm{Hom}_{\mathbf{Z}G}(\partial_1, 1_A)} & \cdots \\
& \downarrow \Phi^{-2} & & \downarrow \Phi^{-1} & & \downarrow \Phi^{0} & & \\
\cdots \to & \mathrm{Hom}_{\mathbf{Z}G}(X_1^*, A) & \xrightarrow{\mathrm{Hom}_{\mathbf{Z}G}(\partial_1^*, 1_A)} & \mathrm{Hom}_{\mathbf{Z}G}(X_0^*, A) & \xrightarrow{\delta^{-1}} & \mathrm{Hom}_{\mathbf{Z}G}(X_0, A) & \xrightarrow{\mathrm{Hom}_{\mathbf{Z}G}(\partial_1, 1_A)} & \cdots
\end{array}$$

where the lower row is the cochain complex obtained from the normalized standard complete resolution of G (see Section 19). Define

$$\Phi^r = 1_{\mathrm{Hom}_{\mathbf{Z}G}(X_r, 1_A)} \qquad \text{for} \quad r = 0, 1, \ldots$$

and Φ^{-r} for $r = 1, 2, \ldots,$ by

$$\Phi^{-r}(a \otimes_{\mathbf{Z}G} [g_1, \ldots, g_{r-1}]) = a * [g_1, \ldots, g_{r-1}]$$

on the generators of $A \otimes_{\mathbf{Z}G} X_{r-1}$. The homomorphism Φ^{-r} is well defined since every element of $A \otimes_{\mathbf{Z}G} X_{r-1}$ can be written uniquely in the form

$$\sum_{g_{\iota_j} \in G} a_\iota \otimes [g_{\iota_1}, \ldots, g_{\iota_{r-1}}]$$

(this is because X_{r-1} is a free $\mathbf{Z}G$-module), and every element of $\mathrm{Hom}_{\mathbf{Z}G}(X_{r-1}^*, A)$ can be written uniquely in the form

$$\sum_{g_{\iota_j} \in G} a_\iota * [g_{\iota_1}, \ldots, g_{\iota_{r-1}}],$$

by Section 21. Hence the vertical maps Φ^r are isomorphisms for every integer r. Moreover $\Phi^0\delta'^{-1} = \delta^{-1}\Phi^{-1}$, for,

$$((\Phi^0\delta'^{-1})a \otimes [\ ])[\ ] = (\sum_{g\in G} f_a^g)[\ ] = Ta$$

$$= \delta^{-1}(a * [\ ])[\ ] \qquad \text{by (23) of Section 21}$$

$$= (\delta^{-1}\Phi^{-1})a \otimes [\ ])[\ ].$$

The verification that the remaining squares are commutative is straightforward. Hence the morphism of complexes $\{\Phi^r\}_{r\in\mathbf{Z}}$ is an isomorphism. Therefore we have the following comparison of the (Tate) cohomology $H^r(G, A)$ of the left G-module A (G finite) with the homology and cohomology groups $H_r^{\mathcal{M}}(G, A)$, $H_{\mathcal{M}}^r(G, A)$ where G is viewed as a monoid:

(30) $$H^{-r}(G, A) \cong H_{r-1}^{\mathcal{M}}(G, A) \qquad \text{for} \quad r = 2, 3, \ldots,$$

(31) $$H^{-1}(G, A) \cong A_T/IA \cong \ker(H_0^{\mathcal{M}}(G, A) \xrightarrow{\delta_*'^{-1}} H_{\mathcal{M}}^0(G, A))$$

where $\delta_*'^{-1}$ is the homomorphism induced by δ'^{-1},

(32) $$H^0(G, A) \cong H_{\mathcal{M}}^0(G, A)/TA \qquad \text{[see 17.2, (23), and (28)]},$$

(33) $$H^r(G, A) \cong H_{\mathcal{M}}^r(G, A) \qquad \text{for} \quad r = 1, 2, \ldots.$$

Thus it is convenient to use these isomorphisms as alternate definitions for the cohomology groups $H^r(G, A)$ for a finite group G (see Serre [93, Chapter 8] and Atiyah-Wall [17, Chapter 6]). In the literature the cohomology groups $H^r(G, A)$ are called the Tate groups and are sometimes denoted by $\hat{H}^r(G, A)$. This notation is used to avoid confusion, since it is customary to denote $H_{\mathcal{M}}^r(G, A)$ by $H^r(G, A)$ when the monoid G is a group. We will continue using the notation $H^r(G, A)$ for the Tate cohomology groups and will point out that $H^0(G, A)$ is A^G whenever G is not finite. When $H^0(G, A)$ is used without any qualification, then it is understood that G is finite and $H^0(G, A) \cong A^G/TA$. As for $r > 0$, we have already shown $H^r(G, A)$ is the same for G finite or infinite [see (33)].

We remark here for use in Chapter 9 that, by 21.2 and (30),

$$H_1(G, A) \cong A \otimes G/[G, G].$$

23. Cohomology of Cyclic Groups

In 14.2 we have shown that if G is a cyclic group of order n generated by g, then

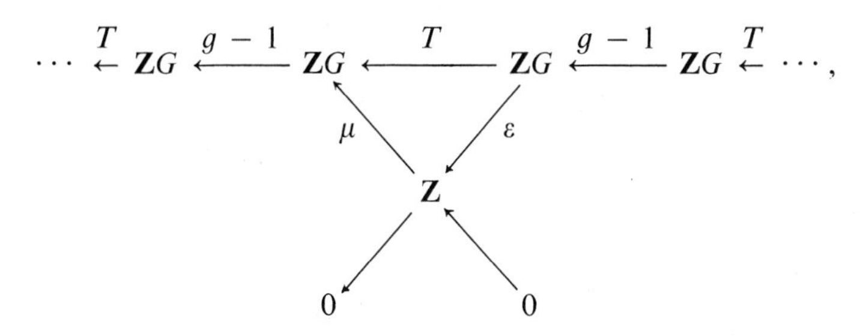

where $\varepsilon(g^i) = 1$ and $\mu(1) = T$, is a complete resolution of G. Since this complex is periodic of period 2, for any G-module M we have

(1) $$H^{2r}(G, M) \cong M^G/TM \quad \text{for} \quad r = 0, \pm 1, \pm 2, \ldots$$

(2) $$H^{2r+1}(G, M) \cong M_T/IM \quad \text{for} \quad r = 0, \pm 1, \pm 2, \ldots$$

by Sections 17.2 and 18.

23.1 *Herbrand's quotient.* Let G be finite cyclic. Let $0 \to L \to M \to N \to 0$ be an exact sequence of G-modules. Then the hexagon

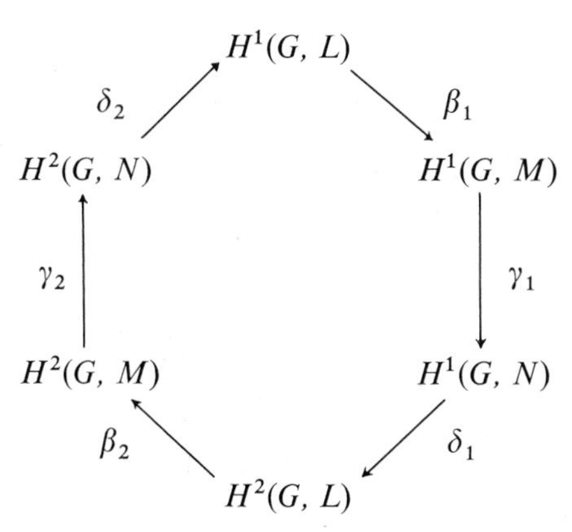

is exact. If the cohomology groups in the hexagon are finite, let $|X|$ denote the order of the group X. Then

$$|H^1(G, L)| = |\delta_2 H^2(G, N)| \cdot |\beta_1 H^1(G, L)|,$$
$$|H^1(G, N)| = |\gamma_1 H^1(G, M)| \cdot |\delta_1 H^1(G, N)|,$$
$$|H^2(G, M)| = |\beta_2 H^2(G, L)| \cdot |\gamma_2 H^2(G, M)|.$$

Hence

(3) $$|H^1(G, L)| \cdot |H^1(G, N)| \cdot |H^2(G, M)| = |H^1(G, M)| \cdot |H^2(G, L)| \cdot |H^2(G, N)|.$$

Define

$$h_{2/1}(M) = \frac{|H^2(G, M)|}{|H^1(G, M)|}.$$

The number $h_{2/1}(M)$ is called *Herbrand's quotient* of the G-module M. By (3) we have:

Let G be a finite cyclic group. Let

$$0 \to L \to M \to N \to 0$$

be an exact sequence of G-homomorphisms. Then if two of the three Herbrand's quotients $h_{2/1}(L)$, $h_{2/1}(M)$, $h_{2/1}(N)$ are defined, so is the third, and moreover

$$h_{2/1}(M) = h_{2/1}(L) \cdot h_{2/1}(N).$$

23.2 *Let G be a finite cyclic group. If M is a finite G-module, then $h_{2/1}(M) = 1$.*

By (1) and (2) of Section 23, $H^2(G, M) = M^G/TM$ and $H^1(G, M) = M_T/IM$. Since $|TM| = |M|/|M_T|$, we have

(4) $$|H^2(G, M)| = |M^G| \cdot |M_T|/|M|.$$

To calculate $|H^1(G, M)|$ we observe that $(g - 1)m = 0$ if and only if $m \in M^G$. Therefore $|IM| = |M|/|M^G|$. Thus

(5) $$|H^1(G, M)| = |M_T| \cdot |M^G|/|M|.$$

By (4) and (5) we have $h_{2/1}(M) = 1$.

23.3 *Let G be a finite cyclic group. Let N be a submodule of the G-module M. If M/N is a finite G-module, then $h_{2/1}(M) = h_{2/1}(N)$.* This follows from 23.1 and 23.2.

23.4 When the finite group G is cyclic of prime order p, Chevalley has shown the following result [18, Theerem 10.3]:

Let G be a cyclic group of prime order p, and let M be a finitely generated G-module. Let n (resp. m) be the rank of M (resp. M^G) as a $\mathbf{Z}$-module. Then

$$h_{2/1}(M)^{p-1} = p^{pm-n}.$$

Tate generalized this result by showing [5, Theorem q. 4, p. XIX]:

Let G be cyclic of prime order p. Let M be a G-module. Let $\varphi(M)$ be Herbrand's quotient if the action of G on M were trivial. If $\varphi(M)$ is defined, then $\varphi(M^G)$ and $h_{2/1}(M)$ exist, and moreover

$$h_{2/1}(M)^{p-1} = \varphi(M^G)^p/\varphi(M). \tag{6}$$

We will show how Chevalley's result follows from (6) after the proof of (6). Let g be a generator of the cyclic group G. Then the sequence of G-homomorphisms

$$0 \longrightarrow M^G \xrightarrow{\;i\;} M \xrightarrow{\;\alpha_{g-1}\;} (g-1)M \longrightarrow 0 \tag{7}$$

is exact, where i is the inclusion map and $\alpha_{\psi(g)}m = \psi(g)m$ for any polynomial $\psi(g)$ in g. Since $(g-1)M$ is a quotient group and a subgroup of M, $\varphi((g-1)M)$ is defined because $\varphi(M)$ is defined. In fact, the composite epimorphism

$$M \xrightarrow{\;\alpha_{g-1}\;} (g-1)M \xrightarrow{\;\pi\;} (g-1)M/p(g-1)M$$

is such that $(\pi\alpha_{g-1})pm = 0$. Thus $\pi\alpha_{g-1}$ induces an epimorphism

$$M/pM \longrightarrow (g-1)M/p(g-1)M.$$

Since M/pM is finite by hypothesis, so is $(g-1)M/p(g-1)M$. Moreover, consider the diagram

$$\begin{array}{ccccc} {}_p(g-1)M & \xrightarrow{\ker \alpha_p'} & (g-1)M & \xrightarrow{\alpha_p'} & (g-1)M \\ \Big\downarrow{\scriptstyle i'} & & \Big\downarrow{\scriptstyle i} & & \\ {}_pM & \xrightarrow{\ker \alpha_p} & M & \xrightarrow{\alpha_p} & M \end{array}$$

where i is the inclusion of $(g-1)M$ and α_p', α_p are multiplications by p. Since

$$\alpha_p \cdot i \cdot \ker \alpha_p' = 0,$$

there exists a unique i': ${}_p(g-1)M \to {}_pM$ such that

$$\ker \alpha_p \cdot i' = i \cdot \ker \alpha_p'.$$

Hence i' is a monomorphism, and so ${}_p(g-1)M$ is finite because ${}_pM$ is finite by hypothesis. We conclude that

$$\varphi((g-1)M) = ((g-1)M : p(g-1)M)/({}_p(g-1)M : 0)$$

is defined. Thus by 23.1, $\varphi(M^G)$ is defined. But $\varphi(M^G) = h_{2/1}(M^G)$, therefore

(8) $$h_{2/1}M = \varphi(M^G)h_{2/1}(g-1)M$$

(9) $$\varphi(M) = \varphi(M^G)\varphi(g-1)M.$$

Substituting (8) and (9) in (6) we get

(10) $$h_{2/1}((g-1)M)^{p-1} = 1/\varphi((g-1)M).$$

Thus we have reduced the problem of proving (6) to the problem of proving (10). Let $(g-1)M = N$. Then $TN = 0$. Thus $N_T = N$, and

(11) $$h_{2/1}(N) = ({}_{(g-1)}N : 0)/(N : (g-1)N)$$

where ${}_{(g-1)}N = N^G$ is the kernel of α_{g-1}. To complete the proof we need the following lemma:

Let N be a G-module. Then

$$({}_{(g-1)^2}N : 0)/(N : (g-1)^2N) = ({}_{(g-1)}N : 0)^2/(N : (g-1)N)^2$$

if either side is defined, where $(X : Y)$ denotes the index of the submodule Y of X in the module X.

PROOF OF THE LEMMA From the exact sequences

$$0 \to {}_{(g-1)}({}_{(g-1)^2}N) \to {}_{(g-1)^2}N \to \alpha_{g-1}({}_{(g-1)^2}N) \to 0$$

$$0 \to \alpha_{g-1}N/\alpha_{(g-1)^2}N \to N/\alpha_{(g-1)^2}N \to N/\alpha_{g-1}N \to 0$$

where ${}_{(g-1)^2}N$ is the kernel of $\alpha_{(g-1)^2}N$, we have the identities

$$({}_{(g-1)^2}N : 0) = (\alpha_{g-1}({}_{(g-1)^2}N) : 0) \cdot ({}_{(g-1)}({}_{(g-1)^2}N) : 0) \tag{12}$$
$$= (\alpha_{g-1}({}_{(g-1)^2}N) : 0) \cdot ({}_{(g-1)}N : 0)$$

and

$$(N : \alpha_{(g-1)^2}N) = (N : \alpha_{g-1}N) \cdot (\alpha_{g-1}N : \alpha_{(g-1)^2}N). \tag{13}$$

Next consider the commutative diagram

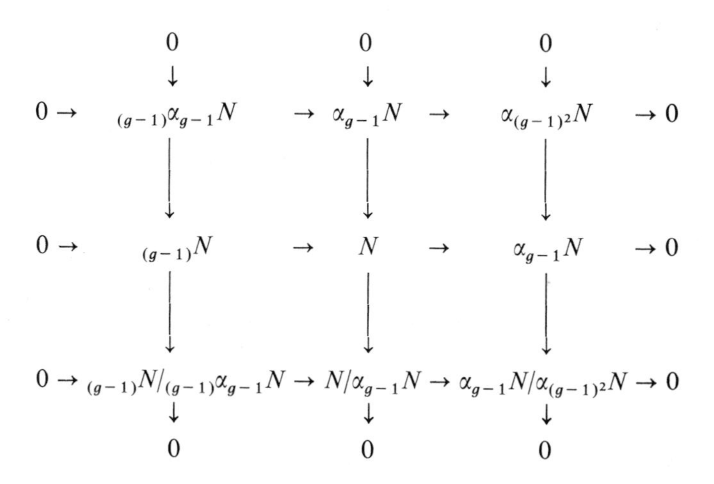

$$\begin{array}{ccccccccc} & & 0 & & 0 & & 0 & & \\ & & \downarrow & & \downarrow & & \downarrow & & \\ 0 & \to & {}_{(g-1)}\alpha_{g-1}N & \to & \alpha_{g-1}N & \to & \alpha_{(g-1)^2}N & \to & 0 \\ & & \downarrow & & \downarrow & & \downarrow & & \\ 0 & \to & {}_{(g-1)}N & \to & N & \to & \alpha_{g-1}N & \to & 0 \\ & & \downarrow & & \downarrow & & \downarrow & & \\ 0 & \to & {}_{(g-1)}N/{}_{(g-1)}\alpha_{g-1}N & \to & N/\alpha_{g-1}N & \to & \alpha_{g-1}N/\alpha_{(g-1)^2}N & \to & 0 \\ & & \downarrow & & \downarrow & & \downarrow & & \\ & & 0 & & 0 & & 0 & & \end{array}$$

where the first two rows and all columns are exact. By the Nine Lemma (see Exercise 2 at the end of Chapter 1), the last row is exact. Therefore we have the identity

$$(N : \alpha_{g-1}N) = (\alpha_{g-1}N : \alpha_{(g-1)^2}N) \cdot ({}_{(g-1)}N : {}_{(g-1)}\alpha_{g-1}N) \tag{14}$$
$$= (\alpha_{g-1}N : \alpha_{(g-1)^2}N) \cdot ({}_{(g-1)}N : \alpha_{g-1}({}_{(g-1)^2}N)).$$

The lemma follows from (12), (13), (14), and the exactness of

$$0 \to \alpha_{g-1}({}_{(g-1)^2}N) \to {}_{(g-1)}N \to {}_{(g-1)}N/\alpha_{g-1}({}_{(g-1)^2}N) \to 0.$$

To complete the proof of Tate's theorem we use (11) and the lemma to get

$$h_{2/1}(N)^{p-1} = ({}_{(g-1)^{p-1}}N : 0)/(N : (g-1)^{p-1}N). \tag{15}$$

Since $TN = 0$, there exists an isomorphism of $\mathbf{Z}G/(T)$ with $\mathbf{Z}[X]/(1 + X + \cdots + X^{p-1})$ the ring of integers in the field of pth roots of unity, i.e., we can treat g as a primitive pth root of unity.

In case p is an odd prime

$$(X - g)(X - g^2) \cdots (X - g^{p-1}) = 1 + X + \cdots + X^{p-1}.$$

So

$$(1 - g)(1 - g^2) \cdots (1 - g^{p-1}) = p.$$

Since $(1 - g^r) = (1 - g)(1 + g + \cdots + g^{r-1})$ we have $(1 - g)^{p-1} | p$. Thus $p = (1 - g)^{p-1} \cdot \varepsilon$ where ε is a unit, so that (15) is the same as

$$h_{2/1}(N)^{p-1} = ({}_pN : 0)/(N : pN) = 1/\varphi(N).$$

In case $p = 2$, $g - 1 = (g + 1) - 2 = T - 2$. Since $TN = 0$, we have ${}_{(g-1)}N = {}_2N$ and $(g - 1)N = 2N$. Hence $h_{2/1}(N) = 1/\varphi(N)$ in this case also.

PROOF OF CHEVALLEY'S THEOREM Since M is finitely generated, $M = M_0 \oplus M_1$ where M_0 is a torsion group and M_1 is torsion free. Then $M^G = M_0^G \oplus M_1^G$. Since M is finitely generated, M_0 is a finite group, and rank M = rank $M_1 = n$, rank M^G = rank $M_1^G = m$. Then

$$h_{2/1}(M)^{p-1} = h_{2/1}(M_1)^{p-1} = \varphi(M_1^G)^p/\varphi(M_1)$$

where $\varphi(M_1^G) = (M_1^G : pM_1^G) = p^m$ and $\varphi(M_1) = (M_1 : pM_1) = p^n$. Hence $h_{2/1}(M)^{p-1} = p^{pm-n}$.

24. Trivial Action

Let G be a finite group. We compute here some cohomology groups in the special case where the action of G on M is trivial, i.e., $gm = m$ for every $g \in G$ and $m \in M$. In what follows the integers $\mathbf{Z}$, the rationals $\mathbf{Q}$, $\mathbf{Q}/\mathbf{Z}$, and M are G-modules on which the action of G are trivial.

(i) $$H^{-2}(G, M) \cong M \otimes G/[G, G] \qquad \text{(see 21.2)}$$

(ii) $$H^{-2}(G, \mathbf{Z}) \cong G/[G, G]$$

and by 15.10,

(iii) $$H^{-3}(G, \mathbf{Q}/\mathbf{Z}) \cong G/[G, G].$$

Since the action of G on M is trivial, $IM = 0$ where I is the kernel of the augmentation map $\varepsilon: \mathbf{Z}G \to \mathbf{Z}$. Moreover $M_T = \{m \in M, \text{ s.t. } Tm = 0\}$ is the kernel of the homomorphism $n \cdot 1_M: M \to M$ where $n = |G|$. Thus, by Section 18,

(iv) $$H^{-1}(G, \mathbf{Z}) = 0,$$

whence, by 15.10,

(v) $$H^{-2}(G, \mathbf{Q}/\mathbf{Z}) = 0.$$

The computation of $H^0(G, M)$ in 17.2 implies

(vi) $$H^0(G, M) \cong M/nM$$

where $n = |G|$. whence

(vii) $$H^0(G, \mathbf{Z}) \cong \mathbf{Z}/n\mathbf{Z},$$

and by 15.10,

(viii) $$H^{-1}(G, \mathbf{Q}/\mathbf{Z}) \cong \mathbf{Z}/n\mathbf{Z}.$$

(ix) $$H^1(G, M) \cong \mathcal{G}(G, M)$$

(see Section 19) where $\mathcal{G}(G, M)$ is the group of homomorphisms from the group G to the abelian group M. In particular,

(x) $$H^1(G, \mathbf{Z}) = 0$$

since G is a finite group. Whence

(xi) $$H^0(G, \mathbf{Q}/\mathbf{Z}) = 0.$$

Furthermore

(xii) $$H^1(G, \mathbf{Q}/\mathbf{Z}) \cong \mathcal{G}(G, \mathbf{Q}/\mathbf{Z}) = \hat{G}$$

where $\hat{G}$ is the group of the characters of G. Thus

$$H^2(G, \mathbf{Z}) \cong \hat{G}. \tag{xiii}$$

We will show in Section 47 that $H^{-r}(G, \mathbf{Z})$ is the dual of $H^r(G, \mathbf{Z})$ [$H^0(G, \mathbf{Z}) = \mathbf{Z}/n\mathbf{Z}$ is self-dual].

CHAPTER

4

Lower Central Series and Dimension Subgroups

25. Dimension Subgroups

This section contains some definitions and elementary results needed in the sequel.

25.1 Let $[G, G]$ be the subgroup of the group G generated by all the elements $\sigma\tau\sigma^{-1}\tau^{-1}$, σ, $\tau \in G$; $[G, G]$ is called the *commutator subgroup* of G. The subgroup $[G, G]$ is normal in G. For, let $x \in [G, G]$, $x = q_1 \cdot \cdots \cdot q_m$, where $q_i = \sigma_i\tau_i\sigma_i^{-1}\tau_i^{-1}$. Then for $\rho \in G$ we have

$$\rho^{-1}x\rho = \rho^{-1}q_1\rho \cdot \rho^{-1}q_2\rho \cdot \cdots \cdot \rho^{-1}q_m\rho$$

where

$$\rho^{-1}q_i\rho = \rho^{-1}\sigma_i\rho \cdot \rho^{-1}\tau_i\rho \cdot (\rho^{-1}\sigma_i\rho)^{-1} \cdot (\rho^{-1}\tau_i\rho)^{-1}$$

is a generator of $[G, G]$. Hence $\rho^{-1}x\rho \in [G, G]$.

25.2 *$G/[G, G]$ is commutative, and satisfies the universal mapping property: Given any homomorphism $f\colon G \to A$ from G to an abelian group A,* *there exists a unique homomorphism $\bar{f}: G/[G, G] \to A$ such that*

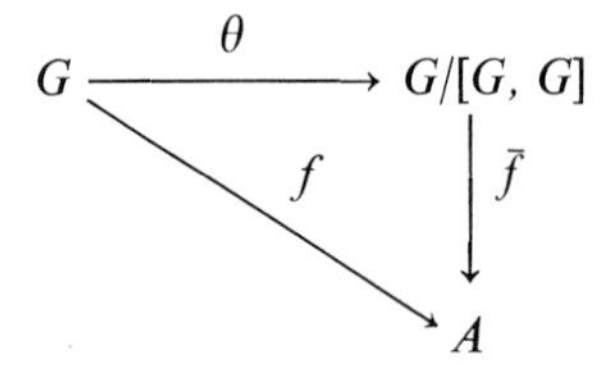

is commutative where θ is the canonical epimorphism from G to the factor group $G/[G, G]$.

The commutativity of $G/[G, G]$ follows from the fact that for any $\sigma, \tau \in G$,

$$\sigma[G, G]\tau[G, G]\sigma^{-1}[G, G]\tau^{-1}[G, G] = \sigma\tau\sigma^{-1}\tau^{-1}[G, G] = [G, G].$$

To show the universal mapping property construct $\bar{f}$ by $\bar{f}(\sigma[G, G]) = f(\sigma)$. Then $\bar{f}\theta = f$. The verification that $\bar{f}$ is well defined and is a homomorphism is straightforward and is left to the reader.

If H, K are subgroups of a group G, denote by $[H, K]$ the subgroup of G generated by the elements $xyx^{-1}y^{-1}$, $x \in H$, $y \in K$.

25.3 *If $U \trianglelefteq G$, then $[U, G] \trianglelefteq G$.*

PROOF For any $\upsilon \in U$ and $\sigma, \tau \in G$ we have

$$\sigma \cdot \upsilon\tau\upsilon^{-1}\tau^{-1} \cdot \sigma^{-1} = \sigma\upsilon\sigma^{-1} \cdot \sigma\tau\sigma^{-1} \cdot (\sigma\upsilon\sigma^{-1})^{-1} \cdot (\sigma\tau\sigma^{-1})^{-1} \in [U, G]$$

since $\sigma\upsilon\sigma^{-1} \in U$. Therefore for any $x \in q_1 \cdot \cdots \cdot q_m \in [U, G]$ where $q_i = \upsilon_i\tau_i\upsilon_i^{-1}\tau_i^{-1}$, $\upsilon_i \in U$, $\tau_i \in G$, we have

$$\sigma x\sigma^{-1} = \sigma q_1\sigma^{-1} \cdot \sigma q_2\sigma^{-1} \cdot \cdots \cdot \sigma q_m\sigma^{-1} \in [U, G].$$

The *lower central series* $\{G_n\}$, $n = 0, 1, \ldots$, of a group is defined by $G_0 = G$, $G_n = [G_{n-1}, G]$ for $n \geqq 1$. If $G_n = \{1\}$ for some n we say G is a *nilpotent group*. If $G_n = \{1\}$ and $G_{n-1} \neq \{1\}$, we say G is *nilpotent of class n*.

We observe that, by 25.1 and 25.3, $G = G_0 \geqq G_1 \geqq G_2 \geqq \cdots$, and $G_i \trianglelefteq G$ for $i = 0, 1, \ldots$.

EXAMPLE Let Z_p be the integers modulo a prime p. Let G be the set of all $n \times n$ matrices of elements of Z_p whose determinants are nonzero; G is a group under the usual multiplication of matrices. The matrix (δ_{ij}), $\delta_{ii} = 1$, $\delta_{ij} = 0$ if $i \neq j$, is the identity element of G. If $(f_{ij}) \in G$, to find the ith row of (g_{ij}), the inverse of the matrix (f_{ij}), solve the system of n equations

$$\sum_{j=1}^{n} g_{ij} f_{jk} = \delta_{ik}, \qquad k = 1, \ldots, n$$

for the unknowns g_{ij}. This system has a unique solution since the determinant of the system is nonzero. We note that the commutator subgroup $[G, G]$ of G is the collection of all matrices whose determinants are equal to 1.

A *descending central series* of a group G is a series $G = U_0 \geqq U_1 \geqq \cdots$ such that $U_n \trianglelefteq G$ and $U_{n+1} \geqq [U_n, G]$.

EXAMPLE Let p be either zero or a prime integer. For each finite group G let $U_0 = G$ and $U_{n+1} = U_n \# G$ for $n \geqq 0$ be the subgroup of G generated by all elements of G of the form $\sigma_n^p \cdot \tau_n \sigma \tau_n^{-1} \sigma^{-1}$ where $\sigma_n, \tau_n \in U_n$, $\sigma \in G$. Then $U_0 \geqq U_1 \geqq \cdots$ is a descending central series, G/U_1 is the maximal quotient group of G which is abelian of exponent p, and for $p \neq 0$ the series is the most rapidly descending central series whose successive quotient groups are vector spaces over the field of p elements. For $p = 0$ the series is the lower central series.

EXAMPLE Let G be the set of all upper triangular $n \times n$ matrices such that each $(\sigma_{ij}) \in G$ satisfies: $\sigma_{ij} = 0$ if $i > j$, $\sigma_{ii} = 1$, σ_{ij} is a real number if $i < j$. If $G_0 \geqq G_1 \geqq \cdots$ is the lower central series of G, then G_1 consists of all $(\sigma_{ij}) \in G$ such that $\sigma_{i,i+1} = 0$ for all i, G_2 consists of all $(\sigma_{ij}) \in G$ such that $\sigma_{i,i+1} = \sigma_{i,i+2} = 0$ for all i, etc.

The lower central series is a descending central series since $G_n \trianglelefteq G$ and $G_n = [G_{n-1}, G]$ for $n = 1, 2, \ldots$. The lower central series is lower than any other descending central series. For, if $G = U_0 \geqq U_1 \geqq \cdots$ is a descending central series, then $G_n \leqq U_n$ for every n, i.e., $G_0 = U_0$, and by induction suppose $G_{n-1} \leqq U_{n-1}$, then

$$G_n = [G_{n-1}, G] \leqq [U_{n-1}, G] \leqq U_n.$$

EXERCISES

1. G nilpotent of class n, $H \leqq G \Rightarrow H$ nilpotent of class less than or equal to n.

2. A homomorphic image of a nilpotent group of class n is nilpotent of class less than or equal to n.

3. If G is nilpotent and H is a maximal proper subgroup, then $H \trianglelefteq G$.

25.4 As in the last chapter, let I be the kernel of the augmentation map $\varepsilon\colon \mathbf{Z}G \to \mathbf{Z}$. The ideal I is a $\mathbf{Z}$-module generated by the family $\{\sigma - 1\}_{\sigma \in G, \sigma \neq 1}$. Define $D_n = \{\sigma \in G \text{ s.t. } \sigma - 1 \in I^{n+1}\}$. It is clear that $G = D_0 \geqq D_1 \geqq \cdots$; D_n is called the nth *dimension subgroup* of G. The series $D_0 \geqq D_1 \geqq \cdots$ is called the dimension series of G. We observe that $D_n \trianglelefteq G$. For, let $\tau \in D_n$, then

$$\tau - 1 = \sum_{i=1}^{k} z_i \prod_{j=1}^{n+1} (\tau_{ij} - 1), \qquad z_i \in \mathbf{Z}, \qquad \tau_{ij} \in G.$$

Thus for $\sigma \in G$,

$$\begin{aligned} \sigma\tau\sigma^{-1} - 1 &= \sigma(\tau - 1)\sigma^{-1} \\ &= \sum_{i=1}^{k} z_i(\sigma\tau_{i,1}\sigma^{-1} - 1)\cdot \cdots \cdot (\sigma\tau_{i,h+1}\,\sigma^{-1} - 1). \end{aligned}$$

Hence $\sigma\tau\sigma^{-1} \in D_n$. Moreover $[D_n, G] \leqq D_{n+1}$; for, let $\sigma \in D_n$, $\tau \in G$, then

$$\begin{aligned} \sigma\tau\sigma^{-1}\tau^{-1} &= 1 + \sigma\tau(\sigma^{-1}\tau^{-1} - \tau^{-1}\sigma^{-1}) \\ &= 1 + \sigma\tau((\sigma^{-1} - 1)(\tau^{-1} - 1) - (\tau^{-1} - 1)(\sigma^{-1} - 1)). \end{aligned}$$

So $\sigma\tau\sigma^{-1}\tau^{-1} \in D_{n+1}$. Hence $[D_n, G] \leqq D_{n+1}$. Therefore *the dimension series is a descending central series*. We conclude that $G_n \leqq D_n$ for every $n = 0, 1, \ldots$. It has been proven that when G is a free group $D_n = G_n$ for every n (Grün [46] and Magnus [72]). The *dimension conjecture* states that $D_n = G_n$ also for an arbitrary group G (Cohn [19]). For $n = 0$ and arbitrary group G we have $D_0 = G = G_0$. For $n = 1$ we have the following:

25.5 *Let G be an arbitrary group, then $G_1 = D_1$.*

PROOF Since $G_1 \leqq D_1$ it is enough to establish the isomorphisms $G/D_1 \cong I/I^2 \cong G/G_1$. To establish the isomorphism $I/I^2 = G/G_1$ define the map $\Phi\colon G \to I/I^2$ by $\Phi\sigma = (\sigma - 1) + I^2$; Φ is a homomorphism since

$$\begin{aligned}\Phi\sigma\tau &= (\sigma\tau - 1) + I^2 \\ &= (\sigma - 1) + (\tau - 1) + (\sigma - 1)(\tau - 1) + I^2 \\ &= (\sigma - 1) + (\tau - 1) + I^2.\end{aligned}$$

Since I/I^2 is abelian, by 25.2, Φ induces the homomorphism $\overline{\Phi}: G/G_1 \to I/I^2$ given by $\overline{\Phi}(\sigma G_1) = (\sigma - 1) + I^2$. Next define the map $\Psi: I \to G/G_1$ by $\Psi(\sigma - 1) = \sigma G_1$; Ψ is a homomorphism because the elements $\sigma - 1$ form a **Z**-basis of I. Since

$$\begin{aligned}\Psi((\sigma - 1)(\tau - 1)) &= \Psi((\sigma\tau - 1) - (\sigma - 1) - (\tau - 1)) \\ &= \sigma\tau G_1 \cdot \sigma^{-1} G_1 \cdot \tau^{-1} G_1 \\ &= \sigma\tau\sigma^{-1}\tau^{-1} G_1 = G_1,\end{aligned}$$

the homomorphism Ψ induces the homomorphism $\overline{\Psi}: I/I^2 \to G/G_1$. It now follows trivially from the definitions that $\overline{\Phi}\overline{\Psi} = 1_{I/I^2}$ and $\overline{\Psi}\overline{\Phi} = 1_{G/G_1}$. Hence $I/I^2 \cong G/G_1$. To establish the isomorphism $G/D_1 \cong I/I^2$ define the map $\chi: I \to G/D_1$ by $\chi(\sigma - 1) = \sigma D_1$; χ is a homomorphism because the elements $\sigma - 1$ form a **Z**-basis of I, and is onto. Moreover

$$\begin{aligned}\chi(\sigma - 1)(\tau - 1) &= \chi(\sigma\tau - 1) - \chi(\sigma - 1) - \chi(\tau - 1) \\ &= \sigma\tau\sigma^{-1}\tau^{-1} D_1 \\ &= D_1\end{aligned}$$

since $G_1 \leqq D_1$. Therefore χ induces the epimorphism $\bar{\chi}: I/I^2 \to G/D_1$. We contend $\bar{\chi}$ is a monomorphism, for, if $\bar{\chi}(\sum_{i=1}^{s} (\sigma_i - 1) + I^2) = D_1$, then $\prod_{i=1}^{s} \sigma_i \in D_1$, and hence

$$\prod_{i=1}^{s} \sigma_i - 1 \in I^2.$$

This together with the formula

$$\sum_{i=1}^{s} (\sigma_i - 1) = (\prod_{i=1}^{s} \sigma_i - 1) - \sum_{j=1}^{s-1} (\sigma_1 \cdot \cdots \cdot \sigma_j - 1)(\sigma_{j+1} - 1)$$

for $s = 2, 3, \ldots,$ easily obtained by induction on s implies that $\sum_{i=1}^{s} (\sigma_i - 1) + I^2 = I^2$. Thus $\bar{\chi}$ is a monomorphism and $I/I^2 \cong G/D_1$ is established.

EXERCISES

1. Using the method in the proof of $[D_n, G] \leqq D_{n+1}$ in 25.4 show $[D_n, D_m] \leqq D_{n+m+1}$ for all $m, n \geqq 0$.
2. Show that $[G_n, G_m] \leqq G_{n+m+1}$ for all $m, n \geqq 0$.
3. Show that if G is nilpotent of class m, then G_n is commutative for $n \geqq (m+1)/2$.

A group G is called *polycyclic* if it has a descending series $G = A_0 \geqq A_1 \geqq \cdots \geqq A_r = 1$ such that $A_i \lhd A_{i-1}$ and A_{i-1}/A_i is cyclic. If moreover $A_i \lhd G$ for every i, then G is called *supersolvable*.

25.6 (P. Hall [48]) *If G is finitely generated and nilpotent, then G is supersolvable.*

PROOF Let $G = \langle a_1, \ldots, a_q \rangle$. Let $G = G_0 \geqq \cdots \geqq G_c = 1$ be the lower central series for G. Denote $[[x_1, x_2], x_3]$ by $[x_1, x_2, x_3]$ and inductively $[[x_1, \ldots, x_k], x_{k+1}]$ by $[x_1, \ldots, x_k]$. For each k there are a finite number of commutators $[a_{i_1}, \ldots, a_{i_k}]$; call these $u_{k_1}, \ldots, u_{k_m}$. Let $G_{k_j} = \langle G_{k+1}, u_{k_1}, \ldots, u_{k_j} \rangle$. For each k interpolate the groups $G_k = G_{k_m} \geqq \cdots \geqq G_{k_1} \geqq G_{k+1}$. We obtain thus a descending central series of G such that each factor group of this series is cyclic.

25.7 (P. Hall [48]) *Let G be a finitely generated nilpotent group, then every subgroup H of G is finitely generated.*

PROOF By 25.6, G is supersolvable. Let $G = A_0 \geqq \cdots \geqq A_r \geqq \{1\}$ be a descending series such that A_{i-1}/A_i is cyclic. Define $H_i = H \cap A_i$. Then

$$H_{i-1}/H_i \cong H_{i-1}/H_{i-1} \cap A_i \cong A_i H_{i-1}/A_i.$$

Hence H_{i-1}/H_i is cyclic, so there is an $x_i \in H_{i-1}$ such that $H_{i-1} = \langle H_i, x_i \rangle$. Thus $H = \langle x_1, \ldots, x_r \rangle$ is finitely generated.

25.8 (Hirsch, cf. [44]) *Let G be polycyclic, then for every $g \in G$, $g \neq 1$, there exists a normal subgroup K_g of finite index in G such that $g \notin K_g$.*

PROOF It is enough to show that if the theorem holds for a polycyclic group H and $G = \langle H, a \rangle$, then the theorem holds for G. Choose $g \neq 1$ in G.

Case 1. $g \notin H$. If G/H is finite let $K_g = H$. If G/H is infinite and $g = a^m h$ for some $h \in H$, let $K_g = \langle H, a^{2m} \rangle$.

Case 2. $g \in H$. By hypothesis there exists M normal and of finite index in H such that $g \notin M$. Let H^m be the subgroup of H generated by all h^m, $h \in G$. Then for large enough m, $M \geqq H^m$. The factor group H/H^m is finite since H is polycylic, hence the automorphism of H/H^m given by $xH^m \to a^{-1}xaH^m$ has finite order, namely, for some $n > 0$, $xa^nH^m = a^nxH^m$. Therefore the subgroup $K_g = \langle H^m, a^n \rangle$ is normal and of finite index in G, and $g \notin K_g$.

25.9 (Gruenberg [44]) *Let G be a finitely generated nilpotent group. If G has an element $g \neq 1$ such that every nontrivial normal subgroup contains g, then G is a finite p-group.*

PROOF Since G is polycyclic, there exists a normal subgroup N of finite index such that $g \notin N$. If G is infinite, N is a nontrivial normal subgroup (of infinite order) and $g \notin N$, contrary to our hypothesis. Therefore, G is finite nilpotent so it can be written as the direct product of its Sylow subgroups. The hypothesis on g allows only one Sylow subgroup in G.

The following result is due to G. Higman [86, Theorem 3.1].

25.10 *If the dimension conjecture is false, then there is a finite p-group for which it is false.*

PROOF Suppose the dimension conjecture false for G. Then there is an $n \geqq 2$ such that $G_n \leqq D_n$, i.e., there exists an $x \in G$ such that $x \notin G_n$ and $x - 1 \in I^{n+1}$. Write

$$x - 1 = \sum_i z_i(x_{i,1} - 1) \cdots (x_{i,n+1} - 1). \tag{1}$$

Let $H \leqq G$ be generated by all elements of G present in (1); H is finitely generated. Let $D_n(H)$ be the nth dimension subgroup of H, and H_n the nth group in the lower central series of H. Then $x \notin H_n$ and $x \in D_n(H)$.

Let $K = H/H_n$; K is finitely generated and nilpotent of class less than or equal to n. Let $k = xH_n$. Then $k \in D_n(K)$ and $k \notin K_n$. Since K is finitely generated and nilpotent, by 25.7, every subgroup of K is finitely generated, and hence by Zorn's lemma there exists a normal subgroup $M \lhd K$ with the property $k \notin M$, and M is maximal with respect to this property. Let $L = K/M$. Let $l = kM$; then $l \notin L_n$ and $l \in D_n(L)$. So the dimension conjecture does not hold in L, and L is finitely generated nilpotent with the element $l \in L$, $l \neq 1$, such that every nontrivial normal subgroup contains l. By 25.9, L is a finite p-group.

26. The Additive Group of Rationals mod 1

Let $\mathbf{T}$ be the additive group of rationals mod 1. We will show $\mathbf{T}$ is *injective* in the category of abelian groups; i.e., given $H \leqq G$, G abelian, $f: H \to \mathbf{T}$ a homomorphism, there exists a homomorphism $g: G \to \mathbf{T}$ such that

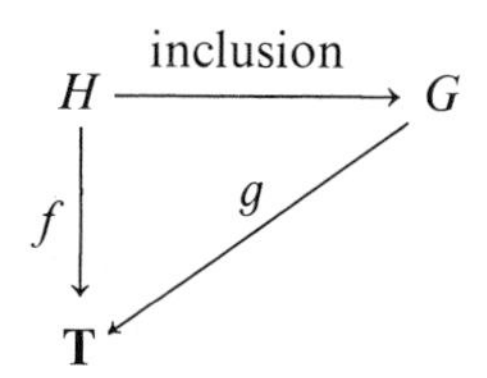

is commutative. An abelian group A is *divisible* if $zA = A$ for every integer $z \neq 0$. Clearly the group $\mathbf{T}$ is divisible. It is also injective because:

26.1 *An abelian group is divisible if and only if it is injective.*

PROOF Let G be divisible. Let $i: A \to B$ be the inclusion of a subgroup A into an abelian group B. Let $\varphi: A \to G$ be a homomorphism. We will show φ can be extended to a homomorphism from B to G. Consider the set of all pairs (A', φ') where the arrows in the diagram

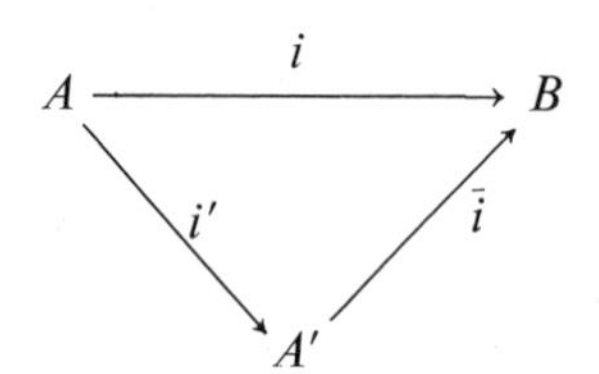

are inclusions, and $\varphi' i' = \varphi$ (i.e., φ' is an extension of φ). Define $(A', \varphi') \leqq (A'', \varphi'')$ if $A' \leqq A'' \leqq B$ and φ'' is an extension of φ'. The set $\mathscr{A}$ of symbols (A', φ') is partially ordered and every totally ordered subset of $\mathscr{A}$ has an upper bound in $\mathscr{A}$. Namely, let $\cdots \leqq (A'_\alpha, \varphi'_\alpha) \leqq \cdots$ be a totally ordered subset. Then $(\bigcup_\alpha A'_\alpha, \varphi)$ is in $\mathscr{A}$ and is an upper bound for the totally ordered subset, where φ is defined as follows: For $a \in \bigcup_\alpha A'_\alpha$, $\varphi(a) = \varphi'_\alpha(a)$ if $a \in A'_\alpha$. The element $\varphi(a)$ is well defined for if $a \in A'_\alpha$, $a \in A'_\beta$, then either $(A'_\alpha, \varphi'_\alpha) \leqq (A'_\beta, \varphi'_\beta)$ or $(A'_\beta, \varphi'_\beta) \leqq (A'_\alpha, \varphi'_\alpha)$; in either case $\varphi(a) = \varphi'_\alpha(a) = \varphi'_\beta(a)$. The symbol $(\bigcup A'_\alpha, \varphi)$ is an upper bound, for, $A'_\gamma \leqq \bigcup_\alpha A'_\alpha$ for every γ, and φ extends each φ'_γ. Hence by Zorn's lemma $\mathscr{A}$ has a maximal element (A_0, φ_0). We contend $A_0 = B$. Namely, if A_0 is a proper subgroup of B, choose $b \in B$ such that $b \notin A_0$, and construct the subgroup $\mathbf{Z}b + A_0$ of B. If $\mathbf{Z}b + A_0$ is a direct sum, then φ_0 can be extended to $\varphi'_0 \colon \mathbf{Z}b + A_0 \to G$ by defining φ'_0 arbitrarily on $\mathbf{Z}b$ and equal to φ_0 on A_0. If $\mathbf{Z}b + A_0$ is not direct, let z be the smallest positive integer such that $zb \in A_0$. Since G is divisible, i.e., $zG = G$, for some $g \in G$ we have $zg = \varphi_0(zb)$. Define $\varphi'_0 \colon \mathbf{Z}b + A_0 \to G$ by

$$\varphi'_0(z'b + a_0) = z'g + \varphi_0(a_0);$$

φ'_0 is well defined, for, if $z''b + a'_0 = z'b + a_0$, then $(z'' - z')b + (a'_0 - a_0) = 0$, so $(z'' - z')b \in A_0$; i.e., $z | z'' - z'$ or $z'' - z' = zz'''$ for some integer z'''. So we have $z'''zb + a'_0 - a_0 = 0$ where both summands are in A'_0. Therefore

$$\begin{aligned}
0 &= \varphi'_0(z'''zb + a'_0 - a_0) \\
&= z'''\varphi_0(zb) + \varphi_0(a'_0 - a_0) \\
&= z'''zg + \varphi_0(a'_0 - a_0) \\
&= (z'' - z')g + \varphi_0(a'_0 - a_0) \\
&= z''g + \varphi_0(a'_0) - (z'g + \varphi_0(a_0)) \\
&= \varphi'_0(z''b + a'_0) - \varphi'_0(z'b + a_0).
\end{aligned}$$

Hence φ'_0 is well defined and extends φ_0, contradicting the maximality of (A_0, φ_0). Therefore $A_0 = B$ and so G is injective.

Conversely, suppose G is injective. Let $z \neq 0$ be an integer. For any $g \in G$ the mapping $\varphi \colon \mathbf{Z}z \to G$ defined by $z'z \to z'g$ is a homomorphism. Extend φ to a homomorphism $\varphi' \colon \mathbf{Z} \to G$ so that

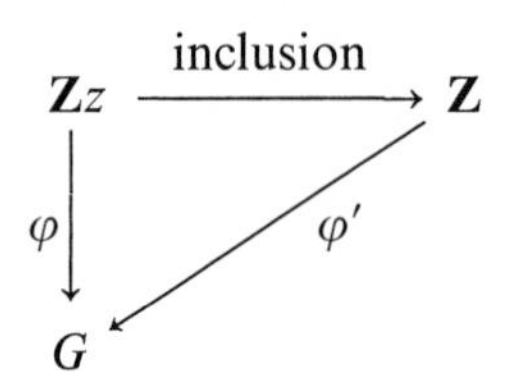

is commutative. Let $g_1 = \varphi'(1) \in G$. Then $\varphi'(z1) = zg_1 = g$. We have shown that for every $g \in G$ there is a $g_1 \in G$ such that $zg_1 = g$; hence $G \subseteq zG$. On the other hand, $zG \subseteq G$ trivially. Therefore $zG = G$ and so G is divisible.

An abelian group C is called a *cogenerator* in the category of abelian groups if given a nontrivial abelian group A and an $a \in A$, $a \neq 1$, there exists a homomorphism $\varphi: A \to C$ such that $\varphi(a) \neq 0$.

26.2 *The group* $\mathbf{T}$, *the additive group of rationals mod* 1, *is a cogenerator in the category of abelian groups.*

PROOF Let A be a nontrivial abelian group and $a \in A$, $a \neq 1$. Let $U = \langle a \rangle$. Let t_a be an arbitrary nonidentity element of $\mathbf{T}$ if $|U|$ is infinite, and $t_a = 1/|U| \in \mathbf{T}$ if $|U|$ is finite. Let $\varphi: U \to \mathbf{T}$ be the homomorphism defined by $\varphi(a) = t_a$. Then φ can be extended to $\varphi_0: A \to \mathbf{T}$ since $\mathbf{T}$ is injective, and $\varphi_0(a) = t_a \neq 1$. Therefore $\mathbf{T}$ is a cogenerator in the category of abelian groups.

We remark that the additive group $\mathbf{Q}^+$ of rationals is injective in the category of abelian groups but it is not a cogenerator since the only homomorphism from any finite group to $\mathbf{Q}^+$ is the trivial homomorphism.

27. Polynomial Maps

Let G be a group, A an abelian group, and φ a mapping of the underlying set of G to A. The mapping φ can be extended by linearity to a homomorphism of the additive group of the group ring $\mathbf{Z}G$ to A. In the sequel φ will denote both the given map φ from G to A and its extension to the homomorphism from $\mathbf{Z}G$ to A. Let I be the augmentation ideal of $\mathbf{Z}G$; then the map $\varphi: G \to A$ is called a *polynomial map of degree less than or equal to n* if φ vanishes on I^{n+1}.

27.1 (*Description of the* nth *dimension group* $D_n(G)$ *of* G *by polynomial maps.*) *Let* G *be a group and* $\mathbf{T}$ *the additive group of rationals* mod 1, *then*

$$D_n(G) = \{g \in G \text{ s.t. } \varphi(g) = \varphi(1) \textit{ for every polynomial map } \varphi\colon G \to \mathbf{T} \textit{ of degree } \leqq n\}.$$

PROOF Let $\varphi\colon G \to \mathbf{T}$ be a polynomial map of degree less than or equal to n. If $g \in D_n(G)$, then $\varphi(g - 1) = 0$ since $g - 1 \in I^{n+1}$. Hence $\varphi(g) = \varphi(1)$. Conversely, let $g \in G$ such that $\varphi(g) = \varphi(1)$ for every polynomial map $\varphi\colon G \to T$ of degree less than or equal to n. We will show $g - 1 \in I^{n+1}$. Suppose $g - 1 \notin I^{n+1}$. Then $g - 1 + I^{n+1}$ is a nonzero element of the abelian group I/I^{n+1}. Since $\mathbf{T}$ is a cogenerator (26.2), there is a homomorphism $\psi\colon I/I^{n+1} \to \mathbf{T}$ such that $\psi(g - 1 + I^{n+1}) \neq 0$. Define $\varphi_0\colon G \to \mathbf{T}$ by $\varphi_0(g) = \psi(g - 1 + I^{n+1})$. We contend φ_0 is a polynomial map of degree less than or equal to n. It is enough to show

$$\varphi_0((g_1 - 1) \cdots (g_k - 1)) = \psi((g_1 - 1) \cdots (g_k - 1) + I^{n+1}). \tag{2}$$

To verify this identity we use induction on k. For $k = 1$,

$$\begin{aligned}\varphi_0(g - 1) = \varphi_0(g) - \varphi_0(1) &= \psi(g - 1 + I^{n+1}) - \psi(1 - 1 + I^{n+1})\\ &= \psi(g - 1 + I^{n+1}).\end{aligned}$$

Suppose by induction that

$$\varphi_0((g_{i_1} - 1) \cdots (g_{i_{k-1}} - 1)) = \psi((g_{i_1} - 1) \cdots (g_{i_{k-1}} - 1) + I^{n+1})$$

for any $g_{i_j} \in G$. Then

$$\begin{aligned}\varphi_0((g_1 - 1) \cdots (g_k - 1)) &= \varphi_0((g_1g_2 - 1)(g_3 - 1) \cdots (g_k - 1))\\ &\quad - \varphi_0((g_1 - 1)(g_3 - 1) \cdots (g_k - 1))\\ &\quad - \varphi_0((g_2 - 1)(g_3 - 1) \cdots (g_k - 1))\\ &= \psi((g_1g_2 - 1)(g_3 - 1) \cdots (g_k - 1) + I^{n+1})\\ &\quad - \psi((g_1 - 1)(g_3 - 1) \cdots (g_k - 1) + I^{n+1})\\ &\quad - \psi((g_2 - 1)(g_3 - 1) \cdots (g_k - 1) + I^{n+1})\\ &= \psi((g_1 - 1)(g_2 - 1) \cdots (g_k - 1) + I^{n+1}).\end{aligned}$$

Hence φ_0 is a polynomial map of degree less than or equal to n with $\varphi_0(g) \neq 0$ and $\varphi_0(1) = 0$. But g was such that $\varphi(g) = \varphi(1)$ for any polynomial map φ of degree less than or equal to n. Therefore $g - 1 \in I^{n+1}$ and so $g \in D_n(G)$.

27.2 *Let G be nilpotent of class n such that $D_{n-1}(G) = G_{n-1}$ where G_{n-1} is the* $(n - 1)$st *term in the lower central series of G. Then $D_n(G) = \{1\}$ if and only if every homomorphism $f\colon G_{n-1} \to \mathbf{T}$ can be extended to a polynomial map $\varphi\colon G \to \mathbf{T}$ of degree less than or equal to n.*

PROOF Suppose $D_n(G) = \{1\}$ and $f\colon G_{n-1} \to \mathbf{T}$ is a homomorphism. Let $\theta\colon G_{n-1} \to I/I^{n+1}$ be the map $g \mapsto g - 1 + I^{n+1}$; θ is a homomorphism since for $g_1, g_2 \in G_{n-1}$,

$$\begin{aligned}\theta(g_1 g_2) &= g_1 g_2 - 1 + I^{n+1}\\ &= (g_1 - 1) + (g_2 - 1) + (g_1 - 1)(g_2 - 1) + I^{n+1}\\ &= \theta(g_1) + \theta(g_2).\end{aligned}$$

θ is monic since $D_n(G) = \{1\}$. Therefore, there exists a homomorphism f' such that

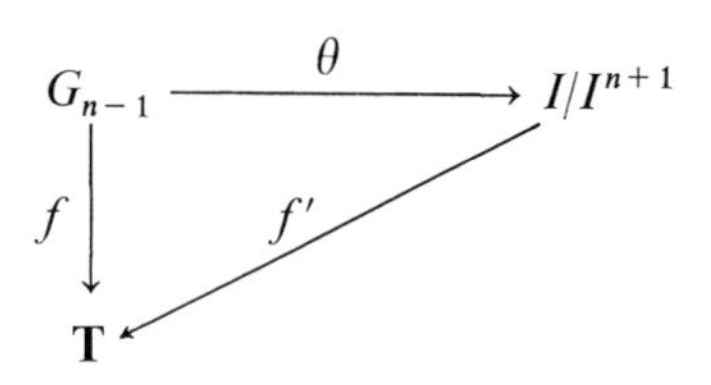

is commutative. Define $\varphi\colon G \to \mathbf{T}$ by $\varphi(g) = f'(g - 1 + I^{n+1})$; φ is a polynomial map of degree less than or equal to n and its restriction to G_{n-1} is f.

Conversely, suppose every homomorphism $f\colon G_{n-1} \to \mathbf{T}$ can be extended to a polynomial map of degree less than or equal to n. Suppose $g \in D_n(G)$ and $g \neq 1$. Then $g \in D_{n-1} = G_{n-1}$. Since G_{n-1} is abelian, there exists a homomorphism $f\colon G_{n-1} \to \mathbf{T}$ such that $f(g) \neq 0$. By hypothesis f can be extended to a polynomial map $\varphi\colon G \to \mathbf{T}$ of degree less than or equal to n and, by 27.1, $\varphi(g) = \varphi(1)$. But $\varphi(1) = f(1) = 0$, contradicting $\varphi(g) = f(g) \neq 0$. Hence $D_n(G) = \{1\}$.

27.3 (Passi [86]) *The dimension conjecture holds for all groups if and only if the property* (P) *holds.*

(P) *If G is a finite p-group of class n, then every homomorphism $f\colon G_{n-1} \to \mathbf{T}$ can be extended to a polynomial map $\varphi\colon G \to \mathbf{T}$ of degree less than or equal to n.*

PROOF Suppose the dimension conjecture holds for all groups. Then if G is a p-group of class n, $f\colon G_{n-1} \to \mathbf{T}$ is a homomorphism; by 27.2, f can be extended to a polynomial map of degree less than or equal to n since $G_n = D_n(G) = \{1\}$. Conversely, suppose the condition (P) holds. If G is a p-group of class 1, the dimension conjecture holds for G since $D_1(G) = G_1$ by 25.5. Suppose the dimension conjecture holds for all p-groups of class less than n. Let G be a finite p-group of class n. Then $D_i(G) = G_i$ for $0 \leqq i < n$. This together with (P), by 27.2, implies $D_n(G) = \{1\} = G_n$. So the dimension conjecture holds for all finite p-groups. By Higman's theorem (25.10) the dimension conjecture holds for all groups.

REMARK In (P) the requirement that G is a finite p-group can be replaced by the condition: G is a finitely generated nilpotent group.

28. Polynomial 2-Cocycles

N. B. In this section all cochains are normalized.

Let $f\colon G \times G \to \mathbf{T}$ be a 2-cochain where the action of G on $\mathbf{T}$ is trivial; f can be extended to a bilinear map $\mathbf{Z}G \times \mathbf{Z}G \to \mathbf{T}$ defined by

$$f\Big(\sum_i z_i g_i, \sum_j z'_j g_j\Big) = \sum_{i,j} z_i z'_j f(g_i, g_j).$$

If f is a 2-cocycle, then

$$\begin{aligned}
(3) \qquad f((g_1 - 1)(g_2 - 1), g_3) &= f(g_1 g_2 - g_1 - g_2 + 1, g_3) \\
&= f(g_1 g_2, g_3) - f(g_1, g_3) - f(g_2, g_3) \\
&= f(g_1, g_2 g_3) - f(g_1, g_2) - f(g_1, g_3) \\
&\qquad \text{since } f \text{ is a 2-cocycle} \\
&= f(g_1, (g_2 - 1)(g_3 - 1)).
\end{aligned}$$

28.1 *Let $f: G \times G \to \mathbf{T}$ be a 2-cocycle. Then*

$$f((g_1 - 1) \cdots (g_k - 1), g_{k+1}) = f(g_1, (g_2 - 1) \cdots (g_{k+1} - 1)) \tag{4}$$

for $k \geqq 2$.

PROOF When $k = 2$ the statement is true by the identity (3) above. Suppose by induction that

$$f((g_{i_1} - 1) \cdots (g_{i_{k-1}} - 1), g_{i_k}) = f(g_{i_1}, (g_{i_2} - 1) \cdots (g_{i_k} - 1))$$

for any $g_{i_j} \in G$. Then

$$\begin{aligned} f((g_1 - 1) &\cdots (g_k - 1), g_{k+1}) \\ &= f((g_1 - 1) \cdots (g_{k-2} - 1)(g_{k-1}g_k - 1), g_{k+1}) \\ &\quad - f((g_1 - 1) \cdots (g_{k-2} - 1)(g_{k-1} - 1), g_{k+1}) \\ &\quad - f((g_1 - 1) \cdots (g_{k-2} - 1)(g_k - 1), g_{k+1}) \\ &= f(g_1, (g_2 - 1) \cdots (g_{k-2} - 1)(g_{k-1}g_k - 1)(g_{k+1} - 1)) \\ &\quad - f(g_1, (g_2 - 1) \cdots (g_{k-2} - 1)(g_{k-1} - 1)(g_{k+1} - 1)) \\ &\quad - f(g_1, (g_2 - 1) \cdots (g_{k-2} - 1)(g_k - 1)(g_{k+1} - 1)) \\ &= f(g_1, (g_2 - 1) \cdots (g_{k+1} - 1)). \end{aligned}$$

We say f is a *polynomial 2-cocycle of degree less than or equal to n* if f is a polynomial map of degree less than or equal to n in the first or second variable. By identity (4) a 2-cocycle is a polynomial map of degree less than or equal to n in either variable if it is a polynomial map of degree less than or equal to n in one of the variables. We say an element $\xi \in H^2(G, \mathbf{T})$ is of degree less than or equal to n if a 2-cocycle representative of ξ is polynomial of degree less than or equal to n.

28.2 *Let $1 \to A \to B \to B/A \to 1$ be exact with A in the center of B. Let U be an abelian group viewed as a trivial B/A-module. Suppose every element of $H^2(B/A, U)$ is of degree less than or equal to n. Then every homomorphism $f: A \to U$ can be extended to a polynomial map $\varphi: B \to U$ of degree less than or equal to $n + 1$. Furthermore if $\rho: B/A \to B$ is a cross section, then ρ can be chosen such that $\varphi(\rho(c)a) = \varphi(\rho(c)) + f(a)$* for all *$c \in B/A$ and $a \in A$.*

PROOF Let $C = B/A$, $\rho: C \to B$ a cross section. Then the elements of B can be written uniquely in the form $\rho(c)\cdot a$. Let $w: C \times C \to A$ be the 2-cocycle obtained from the data

$$1 \to A \to B \underset{\rho}{\rightleftarrows} C \to 1,$$

i.e., $\rho(c_1)\rho(c_2) = \rho(c_1 c_2)w(c_1, c_2)$, and

$$w(c_1, c_2)w(c_1 c_2, c_3) = w(c_1, c_2 c_3)w(c_2, c_3).$$

Let $f: A \to U$ be a homomorphism. Then $fw: C \times C \to U$ is a 2-cocycle where the action of C on U is trivial, i.e.,

$$f(w(c_1, c_2)) + f(w(c_1 c_2, c_3)) = f(w(c_1, c_2 c_3)) + f(w(c_2, c_3)).$$

Let $\xi \in H^2(C, U)$ be the cocycle class of fw. Since ξ is of degree less than or equal to n, there exists a 2-cocycle $k: C \times C \to U$ such that k represents ξ and is a polynomial 2-cocycle of degree less than or equal to n. Since fw and k are cohomologous, there exists a 1-cochain $u: C \to U$, $u(1) = 0$, such that $fw - k = \delta u$. Define $\varphi: B \to U$ by $\varphi(\rho(c)a) = u(c) + f(a)$. Clearly $\varphi(\rho(c)) = u(c)$ and φ extends f. Let $x_1 = \rho(c_1)a_1$, $x_2 = \rho(c_2)a_2$. Then

$$\begin{aligned} \varphi(x_1 x_2) &= \varphi(\rho(c_1)\rho(c_2)a_1 a_2) \\ &= \varphi(\rho(c_1 c_2)w(c_1, c_2)a_1 a_2) \\ &= u(c_1 c_2) + f(w(c_1, c_2)) + f(a_1) + f(a_2). \end{aligned}$$

Hence

$$\begin{aligned} (5) \qquad \varphi((x_1 - 1)(x_2 - 1)) &= \varphi(x_1 x_2) - \varphi(x_1) - \varphi(x_2) \\ &= f(w(c_1, c_2)) - (u(c_2) - u(c_1 c_2) + u(c_1)) \\ &= (fw - \delta u)(c_1, c_2) = k(c_1, c_2). \end{aligned}$$

Using the method of induction in the proof of identity (4) above, the identity (5) implies

$$\varphi((x_1 - 1) \cdots (x_{n+2} - 1)) = k((c_1 - 1) \cdots (c_{n+1} - 1), c_{n+2}) = 0.$$

Hence $\varphi: B \to U$ is polynomial of degree less than or equal to $n + 1$.

The following result is a corollary of 28.2.

28.3 *If G is nilpotent of class n and $H^2(G/G_{n-1}, \mathbf{T})$ is of degree less than or equal to $n-1$, then every homomorphism $f: G_{n-1} \to \mathbf{T}$ can be extended to a polynomial map $\varphi: G \to \mathbf{T}$ of degree less than or equal to n.*

29. Proof of $D_2(G) = G_2$

N. B. In this section all cochains are normalized.

29.1 *Let A, B be abelian groups, $G = A \oplus B$. Then every 2-cocycle $f: G \times G \to \mathbf{T}$ is cohomologous to a 2-cocycle k which satisfies $k(b, a) = 0$, $a \in A$, $b \in B$.*

PROOF Define the map $h: G \to \mathbf{T}$ by $h(ab) = f(b, a)$, $a \in A$, $b \in B$. Then h is a 1-cochain since $h(1) = 0$. Define the 2-cycle $k: G \times G \to \mathbf{T}$ by

$$k(a_1 b_1, a_2 b_2) = f(a_1 b_1, a_2 b_2) + h(a_2 b_2) - h(a_1 a_2 b_1 b_2) + h(a_1 b_1).$$

Then k is cohomologous to f and

$$k(b, a) = f(b, a) + h(a \cdot 1) - h(ab) + h(1 \cdot b) = 0.$$

29.2 *Under the hypothesis of 29.1:*

(i) $$k(a_1 b_1, a_2 b_2) = k(a_1, a_2) + k(a_1, b_2) + k(b_1, b_2)$$

(ii) $$k(a_1 a_2, b) = k(a_1, b) + k(a_2, b)$$

(iii) $$k(a, b_1 b_2) = k(a, b_1) + k(a, b_2).$$

PROOF Throughout the following computations we use identity (4) of 28.1, and the fact that $k(b, a) = 0$ whenever $b \in B$ and $a \in A$.

Proof of (i)

$$\begin{aligned} k(a_1 b_1, a_2 b_2) &= k((a_1 - 1)(b_1 - 1) + (a_1 - 1) \\ &\quad + (b_1 - 1), (a_2 - 1)(b_2 - 1) + (a_2 - 1) + (b_2 - 1)) \\ &= k((a_1 - 1)(b_1 - 1), (a_2 - 1)(b_2 - 1)) \end{aligned}$$

$$
\begin{aligned}
&+ k((a_1 - 1)(b_1 - 1), a_2) \\
&+ k((a_1 - 1)(b_1 - 1), b_2) + k(a_1, (a_2 - 1)(b_2 - 1)) \\
&+ k(a_1, a_2) + k(a_1, b_2) + k(b_1, (a_2 - 1)(b_2 - 1)) \\
&+ k(b_1, a_2) + k(b_1, b_2) \\
= {}& k((b_1 - 1)(b_2 - 1), (a_1 - 1)(a_2 - 1)) \\
&+ k(b_1, (a_1 - 1)(a_2 - 1)) + k((b_1 - 1)(b_2 - 1), a_1) \\
&+ k(b_2, (a_1 - 1)(a_2 - 1)) + k(a_1, a_2) + k(a_1, b_2) \\
&+ k((b_1 - 1)(b_2 - 1), a_2) + k(b_1, b_2) \\
= {}& k(a_1, a_2) + k(a_1, b_2) + k(b_1, b_2).
\end{aligned}
$$

Proof of (ii)

$$
\begin{aligned}
&k(a_1 a_2, b) - k(a_1, b) - k(a_2, b) \\
&\qquad = k((a_1 - 1)(a_2 - 1), b) = k(b, (a_1 - 1)(a_2 - 1)) = 0.
\end{aligned}
$$

Proof of (iii)

$$
\begin{aligned}
&k(a, b_1 b_2) - k(a, b_1) - k(a, b_2) \\
&\qquad = k(a, (b_1 - 1)(b_2 - 1)) = k((b_1 - 1)(b_2 - 1), a) = 0.
\end{aligned}
$$

29.3 *If G is a finite abelian group, then $H^2(G, \mathbf{T})$ is of degree less than or equal to* 1.

PROOF We write G as a direct sum of cyclic groups and use induction on the number of cyclic summands. If G is cyclic of order m, then $H^2(G, \mathbf{T})$ is the subgroup of $\mathbf{T}$ consisting of the elements fixed under the action of G modulo the trace T, where $T = \sum_{g \in G} g$. Since the action of G on $\mathbf{T}$ is trivial $H^2(G, \mathbf{T}) = \mathbf{T}/m\mathbf{T} = 0$ since $m\mathbf{T} = \mathbf{T}$. Hence every 2-cocycle is cohomologous with the trivial cocycle, so $H^2(G, \mathbf{T})$ is of degree less than or equal to 1. Suppose the conclusion true for any finite abelian group which is the direct sum of less than n cyclic groups, $n \geqq 2$, and let G be the direct sum of n cyclic groups. Write $G = A \oplus B$ where A and B are direct sums of less than n cyclic groups. Let $\xi \in H^2(G, \mathbf{T})$. By 29.1 and 29.2 choose a 2-cocycle representative $f\colon G \times G \to \mathbf{T}$ of ξ such that

$$f(a_1b_1, a_2b_2) = f(a_1, a_2) + f(a_1, b_2) + f(b_1, b_2)$$
$$f((a_1 - 1)(a_2 - 1), b) = 0.$$

The restrictions of f to $A \times A$ and $B \times B$ are clearly 2-cocycles; therefore by the induction hypothesis there exist $k_1: A \to \mathbf{T}$, $k_2: B \to \mathbf{T}$ such that

$$u(a_1, a_2) = f(a_1, a_2) - k_1((a_1 - 1)(a_2 - 1))$$
$$v(b_1, b_2) = f(b_1, b_2) - k((b_1 - 1)(b_2 - 1))$$

are polynomial 2-cocycles of degree less than or equal to 1. Define

$$h(a_1b_1, a_2b_2)$$
$$= f(a_1b_1, a_2b_2) - k((a_1 - 1)(a_2 - 1)) - k_2((b_1 - 1)(b_2 - 1)).$$

Then h is a 2-cocycle cohomologous to f on $G \times G$. For, if $w: G \to \mathbf{T}$ is defined by $w(ab) = k_1(a) + k_2(b)$, then

$$(\delta w)(a_1b_1, a_2b_2) = -k_1((a_1 - 1)(a_2 - 1)) - k_2((b_1 - 1)(b_2 - 1)).$$

Furthermore

$$h((a_1b_1 - 1)(a_2b_2 - 1), a_3b_3)$$
$$= f((a_1b_1 - 1)(a_2b_2 - 1), a_3b_3) - k_1((a_1 - 1)(a_2 - 1)(a_3 - 1))$$
$$- k_2((b_1 - 1)(b_2 - 1)(b_3 - 1))$$
$$= f((a_1 - 1)(a_2 - 1), a_3) + f((b_1 - 1)(b_2 - 1), b_3)$$
$$- k_1((a_1 - 1)(a_2 - 1)(a_3 - 1)) - k_2((b_1 - 1)(b_2 - 1)(b_3 - 1))$$
$$= u((a_1 - 1)(a_2 - 1), a_3) + v((b_1 - 1)(b_2 - 1), b_3) = 0.$$

Hence h is a polynomial 2-cocycle of degree less than or equal to 1.

29.4 (Passi [86]) *For any group G, $D_2(G) = G_2$.*

PROOF By Higman's theorem, it is enough to prove the identity for finite p-groups. So, let G be a finite p-group of class 2 and $f: G_1 \to \mathbf{T}$ a homomorphism. Since G/G_1 is a finite abelian group, $H^2(G/G_1, \mathbf{T})$ is of degree less than or equal to 1. Hence, by 28.3, f can be extended to a polynomial map $\varphi: G \to \mathbf{T}$ of degree less than or equal to 2. This, by 27.3, implies $D_2(G) = G_2$.

30. Remarks

Passi [86] has shown that *if G is a group such that none of its subquotients is a 2-group of class 3, then* $D_3(G) = G_3$, *and in case G is such that all its subquotients that are 2-groups of class 3 have order less than or equal to 64, then* $D_3(G) = G_3$. It is also interesting to note that Cohn [19] has shown that the dimension conjecture holds for a group G if $g \in G_n$, $g \notin G_{n+1}$, $g^i \in G_{n+1}$ implies $g^i = 1$.

We have shown in 28.3 that the determination of the higher-dimension subgroups depends on the computation of the degree of $H^2(G, \mathbf{T})$ where G is a finite p-group. The results in the remainder of this section are exploratory in nature. We establish here that $H^2(G, \mathbf{T})$ is of finite degree.

30.1 (Jennings [58]) *Let G be a finite p-group,* I_G *the augmentation ideal of G. Then there exists an integer t such that* $I_G^t \subseteq pI_G$.

PROOF We will use induction on the class of G. Let G be of class 1, i.e., abelian. Let $g_0 = 1, g_1, \ldots, g_{p^r-1}$ be the elements of G. Then $g^{p^r} = 1$ for every $g \in G$, so $(g - 1)^{p^r} \in pI_G$ for every $g \in G$. Since G is abelian I_G^n is generated by the elements $(g_1 - 1)^{n_1} \cdots (g_{p^r-1})^{n_{p^r-1}}$ where $n_1 + \cdots + n_{p^r-1} = n$. These generators of I_G^n form a finite set. Choose n large enough such that at least one of n_i in each generator is greater than or equal to p^r. Then each generator $(g_1 - 1)^{n_1} \cdots (g_i - 1)^{n_i} \cdots (g_{p^r-1})^{n_{p^r-1}} \in pI_G$. Hence for n large enough $I_G^n \subseteq pI_G$. Suppose the assertion true for all finite p-groups of class less than n, and let G be a finite p-group of class n, $n \geqq 2$. Let $(\mathbf{Z}G)I_{G_{n-1}}$ denote the ideal generated by $I_{G_{n-1}}$ in $\mathbf{Z}G$. Then the map

$$\Psi : I_{G/G_{n-1}} \to I_G/(\mathbf{Z}G)I_{G_{n-1}}$$

given by $\Psi(gG_{n-1} - G_{n-1}) = g - 1 + (\mathbf{Z}G)I_{G_{n-1}}$ is well defined, for,

$$\begin{aligned}\Psi(gg_{n-1}G_{n-1} - G_{n-1}) &= gg_{n-1} - 1 + (\mathbf{Z}G)I_{G_{n-1}} \\ &= g - 1 + g(g_{n-1} - 1) + (\mathbf{Z}G)I_{G_{n-1}}\end{aligned}$$

where $g_{n-1} \in G_{n-1}$. Moreover Ψ is an isomorphism. We leave the verification that Ψ is an isomorphism to the reader. By the induction hypothesis for large enough m, $I_{G/G_{n-1}}^m \subseteq pI_{G/G_{n-1}}$. This together with the isomorphism Ψ implies

$$I_G^m/(\mathbf{Z}G)I_{G_{n-1}} \cong I_{G/G_{n-1}}^m \subseteq pI_{G/G_{n-1}} \cong pI_G/(\mathbf{Z}G)I_{G_{n-1}}.$$

Hence

$$I_G^m \subseteq pI_G + (\mathbf{Z}G)I_{G_{n-1}}. \tag{6}$$

Furthermore, by the induction hypothesis, for large enough l, $I_{G_{n-1}}^l \subseteq pI_{G_{n-1}}$. Since G_{n-1} is in the center of G, $((\mathbf{Z}G)I_{G_{n-1}})^l = (\mathbf{Z}G)I_{G_{n-1}}^l$. Hence

$$((\mathbf{Z}G)I_{G_{n-1}})^l \subseteq p(\mathbf{Z}G)I_{G_{n-1}} \subseteq pI_G.$$

This together with (6) implies $I_G^{ml} \subseteq pI_G$.

30.2 *If G is a finite p-group, then $H^2(G, \mathbf{T})$ is of finite degree.*

Proof Let $|G| = p^m$, then $p^m\xi = 0$ for any $\xi \in H^2(G, \mathbf{T})$, by 15.5. Let f: $G \times G \to \mathbf{T}$ be a normalized 2-cocycle representing $\xi \in H^2(G, \mathbf{T})$. Then there exists u: $G \to \mathbf{T}$ such that $u(1) = 0$ and

$$p^m f(g, g') = u(g') - u(gg') + u(g)$$

for every g, $g' \in G$. Define v: $G \to \mathbf{T}$ by $v(g) = p^{-m}u(g)$. Then $v(1) = 0$. Define k: $G \times G \to \mathbf{T}$ by

$$k(g, g') = f(g, g') - v(g') + v(gg') - v(g).$$

Then k is cohomologous to f. Let $a = (g_1 - 1) \cdots (g_t - 1) \in I_G^t$. Then for large enough t, by 30.1, $a = p^m \cdot b$ where $b \in I_G$. Then $k(a, g') = p^m k(b,g')$ for any $g' \in G$. Since, by the definition of k, for g, $g' \in G$ we have

$$p^m k(g, g') = p^m f(g, g') - u(g') + u(gg') - u(g) = 0,$$

it follows that

$$k((g_1 - 1) \cdots (g_t - 1), g') = k(a, g') = p^m k(b, g') = 0 \qquad \text{for every} \quad g' \in G.$$

Hence k: $G \times G \to \mathbf{T}$ is a polynomial 2-cocycle of degree $t - 1$ and represents $\xi \in H^2(G, \mathbf{T})$.

Remark It follows from 28.3 that if a bound is obtained for the integer t above which does not exceed the class of G, then the dimension conjecture follows for all groups.

CHAPTER

5

Relations with Subgroups and Quotient Groups

31. Induced Homomorphisms

31.1 *Induced morphisms of complete resolutions.* Let $\lambda: G' \to G$ be a homomorphism of finite groups. If A is a G-module, we may view A as a G'-module by

$$g'a = (\lambda g')a.$$

Let X_* (resp. X'_*) be a complete resolution of G (resp. G'). View X_* as a complex of G'-modules by means of the homomorphism $\lambda: G' \to G$. We wish to find a chain morphism $\Lambda_*: X'_* \to X_*$ such that the diagram

(1)

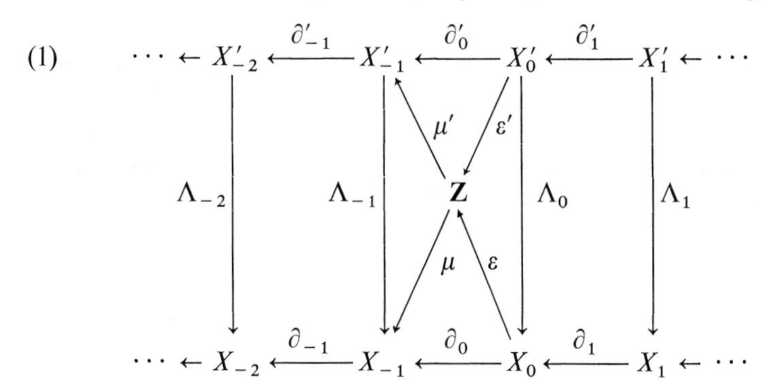

is commutative. To construct Λ_0 we observe that in the diagram

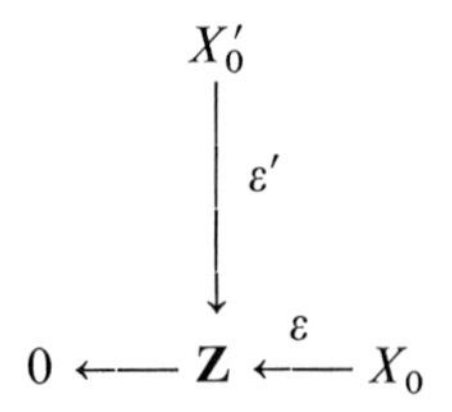

the row is exact and X_0' is G'-free, therefore G'-projective; thus there exists a G'-homomorphism $\Lambda_0 \colon X_0' \to X_0$ such that

$$\varepsilon\Lambda_0 = \varepsilon'.$$

We point out that Λ_0 is not unique. To construct Λ_1 we observe that $\varepsilon(\Lambda_0\partial_1') = 0$. Thus by the exactness of the lower row in (1) at X_0 the image of X_1' under $\Lambda_0\partial_1'$ is contained in the image of ∂_1. Therefore again we have

$$\begin{array}{ccccc} & & X_1' & & \\ & & \downarrow & & \\ 0 & \longleftarrow & \operatorname{im}\partial_1 & \longleftarrow & X_1 \end{array}$$

where the row is exact and X_1' is G'-projective. Hence there exists a G'-homomorphism $\Lambda_1 \colon X_1' \to X_1$ such that

$$\partial_1\Lambda_1 = \Lambda_0\partial_1'.$$

Similarly we can construct $\Lambda_2, \Lambda_3, \ldots$, such that

$$\partial_r\Lambda_r = \Lambda_{r-1}\partial_r', \qquad r = 1, 2, \ldots.$$

To construct Λ_r for $r < 0$, X_r for $r < 0$ must be G'-free. [X_r is G-free but is not G'-free unless λ is a monomorphism. For, if $l \in \ker \lambda$, $l \neq 1$, then for any $\xi \in X_r$, $\xi \neq 0$, we have $(l - 1)\xi = 0$.] In the rest of this section we will assume λ is a monomorphism. Thus X_r is a G'-free module for every integer r. By 13.1, X_r is G'-induced; hence there exists a **Z**-homomorphism $\pi_r \in \operatorname{Hom}(X_r, X_r)$ such that $T'(\pi_r) = 1_{X_r}$ (13.2), where $T' = \sum_{g' \in G'} g'$. For $r = 0, 1, 2, \ldots$, define

$$\Lambda_{r-1} = T'(\pi_{r-1}\partial_r\Lambda_r D_{r-1}')$$

where the **Z**-homomorphisms D'_r exist by 9.4. Suppose $\partial_{r+1}\Lambda_{r+1} = \Lambda_r\partial'_{r+1}$, then

$$\begin{aligned}\Lambda_{r-1}\partial'_r &= T'(\pi_{r-1}\partial_r\Lambda_r D'_{r-1}\partial'_r) \qquad \text{since} \quad \partial'_r \text{ is a } G'\text{-homomorphism}\\ &= T'(\pi_{r-1}\partial_r\Lambda_r) - T'(\pi_{r-1}\partial_r\Lambda_r\partial'_{r+1}D'_r)\\ &= T'(\pi_{r-1}\partial_r\Lambda_r) - T'(\pi_{r-1}\partial_r\partial_{r+1}\Lambda_{r+1}D'_r)\\ &= T'(\pi_{r-1})\partial_r\Lambda_r \qquad \text{since} \quad \partial_r\Lambda_r \text{ is a } G'\text{-homomorphism}\\ &= \partial_r\Lambda_r.\end{aligned}$$

Since $\partial_1\Lambda_1 = \Lambda_0\partial'_1$, by induction we have

$$\partial_r\Lambda_r = \Lambda_{r-1}\partial'_r \qquad \text{for} \quad r < 0.$$

To complete the proof of the commutativity of the diagram (1) we must show $\Lambda_{-1}\mu' = \mu$. This follows from the identities $\partial_0\Lambda_0 = \Lambda_{-1}\partial'_0$, $\varepsilon\Lambda_0 = \varepsilon'$, $\mu'\varepsilon' = \partial'_0$, $\mu\varepsilon = \partial_0$, and the fact that ε' is an epimorphism.

31.2 *Any two chain morphisms* $\Lambda_*, \Lambda'_*: X'_* \rightrightarrows X_*$ *induced by the monomorphism* $\lambda: G \to G'$ *are homotopic.* The chain morphism $\Lambda_*: X'_* \to X_*$ is not uniquely constructed. Let $\Lambda_*, \Lambda'_*: X'_* \rightrightarrows X_*$ be two chain morphisms induced by the monomorphism $\lambda: G' \to G$. Let

$$M_* = \Lambda_* - \Lambda'_*.$$

Then

$$M_r\partial'_{r+1} = \partial_{r+1}M_{r+1} \tag{2}$$

and

$$\varepsilon M_0 = 0. \tag{3}$$

We claim $M_*: X'_* \to X_*$ is homotopic to zero. For, consider

$$\begin{array}{ccccc} & & X'_0 & & \\ & & \downarrow{\scriptstyle M_0} & & \\ \mathbf{Z} & \xleftarrow{\varepsilon} & X_0 & \xleftarrow{\partial_1} & X_1. \end{array} \tag{4}$$

By (3), the diagram (4) implies

$$\begin{array}{ccccc} & & X_0' & & \\ & & \downarrow M_0 & & \\ 0 & \longleftarrow & \operatorname{im} \partial_1 & \xleftarrow{\partial_1} & X_1 \end{array}$$

where the row is exact. Since X_0' is G'-free, there exists $\Delta_0 \colon X_0' \to X_1$ such that

$$\partial_1 \Delta_0 = M_0. \tag{5}$$

Next consider

$$\begin{array}{ccccc} & & X_1' & & \\ & & \downarrow M_1 - \Delta_0 \partial_1' & & \\ X_0 & \xleftarrow{\partial_1} & X_1 & \xleftarrow{\partial_2} & X_2. \end{array}$$

Since

$$\begin{aligned} \partial_1(M_1 - \Delta_0 \partial_1') &= \partial_1 M_1 - M_0 \partial_1' \qquad \text{by (5)} \\ &= 0, \end{aligned}$$

we have

$$\begin{array}{ccccc} & & X_1' & & \\ & & \downarrow M_1 - \Delta_0 \partial_1' & & \\ 0 & \longleftarrow & \operatorname{im} \partial_2 & \xleftarrow{\partial_2} & X_2 \end{array}$$

where the row is exact. Since X_r' is G'-free, there exists $\Delta_1 \colon X_1' \to X_2$ such that

$$\partial_2 \Delta_1 = M_1 - \Delta_0 \partial_1'.$$

Suppose $\Delta_0, \ldots, \Delta_r$ are defined and $\partial_{r+1} \Delta_r + \Delta_{r-1} \partial_r' = M_r$. Since

$$\partial_{r+1}(M_{r+1} - \Delta_r \partial_{r+1}') = \partial_{r+1} M_{r+1} - M_r \partial_{r+1}' + \Delta_{r-1} \partial_r' \partial_{r+1}' = 0,$$

we have

$$\begin{array}{ccc} & & X'_{r+1} \\ & & \Big\downarrow M_{r+1} - \Delta_r \partial'_{r+1} \\ 0 \longleftarrow & \operatorname{im} \partial_{r+2} & \overset{\partial_{r+2}}{\longleftarrow} X_{r+2} \end{array}$$

where the row is exact and X'_{r+1} is G'-free. Therefore there exists a G'-homomorphism $\Delta_{r+1}: X'_{r+1} \to X_{r+2}$ such that $\partial_{r+2}\Delta_{r+1} = M_{r+1} - \Delta_r\partial'_{r+1}$. Thus M_r for $r > 1$ is homotopic to zero. Define $\Delta_{-1} = 0$; then M_0 is also homotopic to zero. Define $\Delta_{r-2}: X'_{r-2} \to X_{r-1}$, $r \leqq 0$, by

$$\Delta_{r-2} = T'(\pi_{r-1}(M_{r-1} - \partial_r\Delta_{r-1})D'_{r-2}).$$

Then

$$\begin{aligned} \Delta'_{-2}\partial'_{-1} &= T'(\pi_{-1}M_{-1}D'_{-2}\partial'_{-1}) \\ &= T'(\pi_{-1}M_{-1}(1_{X_{-1}} - \partial'_0 D_{-1})) \\ &= T'(\pi_{-1}M_{-1} - T'(\pi_{-1}\partial_0 M_0 D_{-1}) \\ &= M_{-1} - T'(\pi_{-1}\partial_0\partial_1\Delta_0 D_{-1}) \\ &= M_{-1} \end{aligned}$$

and

$$\begin{aligned} &\Delta_{r-2}\partial'_{r-1} \\ &\quad = T'(\pi_{r-1}(M_{r-1}\partial_r\Delta_{r-1})D'_{r-2}\partial'_{r-1}) \\ &\quad = T'(\pi_{r-1}(M_{r-1} - \partial_r\Delta_{r-1})) - T'(\pi_{r-1}(M_{r-1} - \partial_r\Delta_{r-1})\partial'_r D'_{r-1}) \\ &\quad = T'(\pi_{r-1}(M_{r-1} - \partial_r\Delta_{r-1}) - T'(\pi_{r-1}(\partial_r M_r D'_{r-1} - \partial_r\Delta_{r-1}\partial'_r D'_{r-1})) \\ &\quad = T'(\pi_{r-1})(M_{r-1} - \partial_r\Delta_{r-1}) \qquad \text{since } \Delta_{r-1} \text{ is a } G'\text{-homomorphism} \\ &\quad = M_{r-1} - \partial_r\Delta_{r-1}. \end{aligned}$$

Thus M_* is homotopic to zero. This implies that any two chain morphisms $\Lambda_*, \Lambda'_*: X'_* \rightrightarrows X_*$ induced by the monomorphism $\lambda: G \to G'$ are homotopic.

31.3 *Induced homomorphisms of cohomology groups.* In this section we consider objects (G, A) where G is a finite group and A is a G-module. We say

$$(\lambda, f): (G, A) \to (G', A')$$

is a morphism from the object (G, A) to the object (G', A') if $\lambda: G' \to G$ is a monomorphism and $f: A \to A'$ is a homomorphism of G'-modules. [It is clear that we are considering the G'-module structure of A by means of λ, $f(\lambda(g')a) = g'f(a)$.]

Let A_i be a G_i-module, $i = 1, 2, 3$. If $\lambda_j: G_{j+1} \to G_j$, $j = 1, 2$, is a monomorphism, and $f_j: A_j \to A_{j+1}$, $j = 1, 2$, is a G_{j+1}-homomorphism, we define the composition of morphisms

$$(\lambda_2, f_2)(\lambda_1, f_1) = (\lambda_1\lambda_2, f_2 f_1).$$

Then $(1_G, 1_A) = 1_{(G,A)}$. [The objects (G, A) and the morphisms (λ, f) described above form a category.]

Construct the homomorphism

$$(\lambda, f)_r^*: H^r(G, A) \to H^r(G', A')$$

as follows: If t is an r-cochain, $t \in \mathrm{Hom}_G(X_r, A)$, define

$$(\lambda, f)_r'(t) = ft\Lambda_r \in \mathrm{Hom}_{G'}(X_r', A').$$

This is possible because a G-homomorphism is a G'-homomorphism, i.e.,

$$t(g'x) = t(\lambda g' \cdot x) = (\lambda g')(tx) = g' \cdot t(x).$$

When t is an r-cocycle, $t\partial_{r+1} = 0$. Thus $ft\Lambda_r\partial_{r+1}' = ft\partial_{r+1}\Lambda_{r+1} = 0$. Therefore the image of an r-cocycle under $(\lambda, f)_r'$ is an r-cocycle. If t is a coboundary, then $t = s\partial_r$ where $s \in \mathrm{Hom}_G(X_{r-1}, A)$. Thus $(\lambda, f)_r't = fs\partial_r\Lambda_r = fs\Lambda_{r-1}\partial_r'$. Therefore $(\lambda, f)_r'$ carries coboundaries to coboundaries. Hence (λ, f) induces a homomorphism

$$(\lambda, f)_r^*: H^r(G, A) \to H^r(G', A').$$

For the moment, $(\lambda, f)_r^*$ depends on λ and the complete resolutions X_*, X_*' of G and G', respectively, and is independent of the morphism $\Lambda_*: X_*' \to X_*$ by 31.2. To show the independence of $(\lambda, f)_r^*$ from X_* and X_*' we first observe that the composite morphism

$$(G, A) \xrightarrow{(\lambda, f)} (G', A') \xrightarrow{(\lambda', f')} (G'', A'')$$

induces the composite homomorphism

$$(6)\qquad (\lambda', f')^*_{(X_*', X_*'')} \cdot (\lambda, f)^*_{(X_*, X_*')} = (\lambda\lambda', f'f)^*_{(X_*, X_*'')}.$$

In fact the monomorphism $\lambda\lambda'$ induces the chain morphism $\Lambda_*\Lambda'_*\colon X''_* \to X_*$, for

$$\text{(i)}\qquad \varepsilon\Lambda_0\Lambda'_0 = \varepsilon'\Lambda'_0 = \varepsilon'',$$

$$\text{(ii)}\qquad \Lambda_r\Lambda'_r\partial''_{r+1} = \Lambda_r\partial'_{r+1}\Lambda'_{r+1} = \partial_{r+1}\Lambda_{r+1}\Lambda'_{r+1}.$$

Thus both sides of (6) carry the class of an r-cocycle t to the class of $f'ft\Lambda_r\Lambda'_r$. Moreover, if $G' = G$, $A' = A$, $\lambda = 1_G$, and $f = 1_A$, we may still have different complete resolutions X_* and X'_* of G. Denote $(1, 1)^*_{(X_*, X_*')}$ by $1_{X_*, X_*'}$ and $(1, 1)^*_{X_*', X_*}$ by $1_{X_*', X_*}$. Then, by (6),

$$1_{X_*, X_*'} \cdot 1_{X_*', X_*} = 1_{X_*', X_*'} \qquad \text{and} \qquad 1_{X_*', X_*} \cdot 1_{X_*, X_*'} = 1_{X_*, X_*}$$

Since for $1_{X_*, X_*}$ we may use $\Lambda_* = 1_*$ (by 31.2) we conclude that $1_{X_*, X_*}$ is identity. Therefore if $(G, A)_{X_*}$ stands for the cochain complex with coefficients in the G-module A obtained from the complete resolution X_* of G [see 14.3,(2)], the diagram

$$\begin{array}{ccc} (G, A)_{X_*} & \xrightarrow{(\lambda, f)'} & (G', A')_{X_*'} \\ \downarrow{\scriptstyle 1_{X_*, \bar{X}_*}} & & \downarrow{\scriptstyle 1_{X_*', \bar{X}_*'}} \\ (G, A)_{\bar{X}_*} & \xrightarrow{(\lambda, f)'} & (G', A')_{\bar{X}_*'} \end{array}$$

is commutative up to homotopy by 31.2. Hence

$$\begin{array}{ccc} H^r_{X_*}(G, A) & \xrightarrow{(\lambda, f)^*_r} & H^r_{X_*'}(G', A') \\ \downarrow{\scriptstyle 1_{X_*, \bar{X}_*}} & & \downarrow{\scriptstyle 1^*_{X_*', \bar{X}_*'}} \\ H^r_{\bar{X}_*}(G, A) & \xrightarrow{(\lambda, f)^*_r} & H^r_{\bar{X}_*'}(G', A') \end{array}$$

is commutative. Thus $(\lambda, f)_r^*$ is independent of the complete resolutions X_* and X'_* for G and G', respectively. [Indeed we have shown that the assignment of $H^r(G, A)$ to the pair (G, A) is a functor from the category whose objects are the pairs (G, A) and whose morphisms are the (λ, f) to the category of abelian groups.]

31.4 $\lambda: G' \to G$ as before is a monomorphism and

$$\begin{array}{ccccccccc} 0 \to & A & \xrightarrow{i} & B & \xrightarrow{j} & C & \to 0 \\ & f\big\downarrow & & g\big\downarrow & & h\big\downarrow & \\ 0 \to & A' & \xrightarrow{i'} & B' & \xrightarrow{j'} & C' & \to 0 \end{array}$$

is a commutative diagram where the upper row is an exact sequence of G-modules, the lower row is an exact sequence of G'-modules, and f, g, h are G'-homomorphisms. Let X_* be a complete resolution of G and X'_* a complete resolution of G'. Then the diagram of morphisms of cochain complexes

$$\begin{array}{ccccccc} 0 \to \mathrm{Hom}_G(X_*, A) & \xrightarrow{\mathrm{Hom}_G(1_{X_*}, i)} & \mathrm{Hom}_G(X_*, B) & \xrightarrow{\mathrm{Hom}_G(1_{X_*}, j)} & \mathrm{Hom}_G(X_*, C) \to 0 \\ (\Lambda_*, f)'\big\downarrow & & (\Lambda_*, g)'\big\downarrow & & (\Lambda_*, h)'\big\downarrow \\ 0 \to \mathrm{Hom}_{G'}(X'_*, A') & \xrightarrow{\mathrm{Hom}_{G'}(1_{X'_*}, i)} & \mathrm{Hom}_{G'}(X', B') & \xrightarrow{\mathrm{Hom}_{G'}(1_{X'_*}, j)} & \mathrm{Hom}_{G'}(X'_*, C') \to 0 \end{array}$$

is commutative (by diagram chasing), moreover the rows are exact by 7.7. It follows by 8.5 that

$$\begin{array}{ccccccccc} \cdots \to & H^r(G, B) & \to & H^r(G, C) & \xrightarrow{\delta^*} & H^{r+1}(G, A) & \to & H^{r+1}(G, B) & \to \cdots \\ & (\lambda, g)_r^*\big\downarrow & & (\lambda, h)_r^*\big\downarrow & & (\lambda, f)_{r+1}^*\big\downarrow & & (\lambda, g)_{r+1}^*\big\downarrow & \\ \cdots \to & H^r(G', B') & \to & H^r(G', C') & \xrightarrow{\delta'^*} & H^{r+1}(G', A') & \to & H^{r+1}(G', B') & \to \cdots \end{array}$$

is commutative.

32. The Restriction and Corestriction (or Transfer) Maps

32.1 Let G be a finite group and $\lambda: S \to G$ the inclusion in G of the subgroup S of G. Let M be a G-module, then M is an S-module as well. The homomorphism $(\lambda, 1_M)_r^*$ of Section 31 is denoted by $\mathrm{res}_{G,S}^r$: $H^r(G, M) \to H^r(S, M)$ (or simply $\mathrm{res}_{G,S}$) and is called the *restriction* map from G to S.

32.2 Let $S \leqq G$, $\lambda: S \to G$ the inclusion of S in G. Let X_*(resp. X_*') be a complete resolution of G (resp. S). By 31.1, λ induces a chain morphism $\Lambda_*': X_*' \to X_*$. Similarly we construct a morphism $\Lambda_*: X_* \to X_*'$ of complexes of S-modules such that the diagram

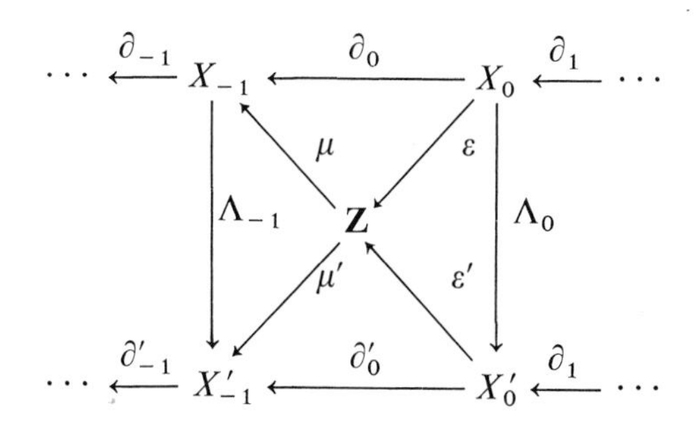

is commutative. If M is a G-module (therefore also an S-module), define the cochain morphism *corestriction* or *transfer* $\mathrm{Hom}_S(X_*', M) \to \mathrm{Hom}_G(X_*, M)$ by

$$\mathrm{cor}_{S,G}^r(t') = T_{S,G}(t'\Lambda_r) = \sum_{i=1}^{n} g_i t' \Lambda_r g_i^{-1}$$

where $G = \dot{\bigcup}_{i=1}^{n} g_i S$. We note that

$$\begin{aligned}\delta^r(\mathrm{cor}_{S,G}^r(t')) &= T_{S,G}(t'\Lambda_r \partial_{r+1}) \\ &= T_{S,G}(t'\partial'_{r+1}\Lambda_{r+1}) \\ &= T_{S,G}(\delta'^r(t')\Lambda_{r+1}) \\ &= \mathrm{cor}_{S,G}^{r+1}(\delta'^r(t')).\end{aligned}$$

Thus cor maps coboundaries to coboundaries and cocycles to cocycles. Hence $\mathrm{cor}^*_{S,G}\colon \mathrm{Hom}_S(X'_*, M) \to \mathrm{Hom}_G(X_*, M)$ induces the homomorphism

$$\mathrm{cor}^r_{S,G}\colon H^r(S, M) \to H^r(G, M)$$

such that

$$\mathrm{cor}^r_{S,G}(t'_r + \delta' \,\mathrm{Hom}_S(X'_{r+1}, M)) = \mathrm{cor}^r_{S,G}(t'_r) + \delta \,\mathrm{Hom}_G(X_{r+1}, M).$$

We remark that if $X'_* = X_*$, and $\Lambda_* = 1_{*,S}$ where $1_{*,S}$ is the identity morphism of the complex X_* viewed as a complete resolution of S, then

$$\mathrm{cor}^r_{S,G}(t) = T_{S,G}(t) = \sum_{i=1}^{n} g_i t g_i^{-1} \qquad \text{where} \quad G = \bigcup_{i=1}^{n} g_i S.$$

32.3 *Let $\lambda\colon G' \to G$ be a monomorphism, M (resp. M') a G-(resp. G') module. Let $f\colon M \to M'$ be a G'-homomorphism, $S \leqq G$, $S' \leqq G'$. If*

(i) $\lambda(S') \leqq S$

(ii) $\lambda(G')S = S\lambda(G') = G$

(iii) $(G'\colon S') = (G\colon S)$,

then the diagram

$$\begin{array}{ccc} H^r(S, M) & \xrightarrow{(\lambda, f)^*} & H^r(S', M) \\ \downarrow {\scriptstyle \mathrm{cor}^r_{S,G}} & & \downarrow {\scriptstyle \mathrm{cor}^r_{S',G'}} \\ H^r(G, M) & \xrightarrow{(\lambda, f)^*} & H^r(G', M) \end{array} \tag{7}$$

is commutative.

PROOF Let $c' = S'\bar{c}'$ be a left coset of S' in G'. Let $G' = \dot{\bigcup}_{c'} S'\bar{c}'$. Then $\lambda(G') = \bigcup_{c'} \lambda S' \cdot \lambda \bar{c}'$. By (i) and (ii) we have

$$G = \dot{\bigcup_{c'}} S \cdot \lambda \bar{c}',$$

and this is a disjoint union by (iii). To show the diagram (7) is commutative, on one hand

$$(\lambda, f)^* \mathrm{cor}^r_{S,G}(t) = \sum_{c'} \bar{c}' f t \Lambda_r \bar{c}'^{-1} = T_{S',G'}(f t \Lambda_r)$$

since $f\lambda(g') = g' \cdot f$ and $\lambda(\bar{c}')^{-1}\Lambda_r = \Lambda_r(\bar{c}')^{-1}$, and on the other hand

$$\mathrm{cor}^r_{S',G'}(\lambda, f)^*(t) = \mathrm{cor}^r_{S',G'}(f t \Lambda_r) = T_{S',G'}(f t \Lambda_r).$$

32.4 *If $S \leqq G$ and*

$$0 \to L \xrightarrow{i} M \xrightarrow{j} N \to 0$$

is an exact sequence of G-modules (therefore of S-modules), then

$$\begin{array}{ccc} H^r(S, N) & \xrightarrow{\delta^*} & H^{r+1}(S, L) \\ \downarrow \mathrm{cor}^r_{S,G} & & \downarrow \mathrm{cor}^r_{S,G} \\ H^r(G, N) & \xrightarrow{\delta^*} & H^{r+1}(G, L) \end{array}$$

is commutative.

PROOF Let X_* be a complete resolution of G, then it may be viewed as a complete resolution of S. Consider the diagram of cochain complexes

(8)

$$\begin{array}{ccccc} 0 \to \mathrm{Hom}_S(X_*, L) & \xrightarrow{\mathrm{Hom}_S(1_{X_*}, i)} & \mathrm{Hom}_S(X_*, M) & \xrightarrow{\mathrm{Hom}_S(1_{X_*}, j)} & \mathrm{Hom}_S(X_*, N) \to 0 \\ \downarrow \mathrm{cor}^*_{S,G} & & \downarrow \mathrm{cor}^*_{S,G} & & \downarrow \mathrm{cor}^*_{S,G} \\ 0 \to \mathrm{Hom}_G(X_*, L) & \xrightarrow{\mathrm{Hom}_G(1_{X_*}, i)} & \mathrm{Hom}_G(X_*, M) & \xrightarrow{\mathrm{Hom}_G(1_{X_*}, j)} & \mathrm{Hom}_G(X_*, N) \to 0 \end{array}$$

where the rows are exact in each degree. The diagram (8) is commutative since

$$\mathrm{cor}^r_{S,G}\,\mathrm{Hom}_S(1_{X_r}, i)t = T_{S,G}(it) = iT_{S,G}(t) = \mathrm{Hom}_G(1_{X_r}, i)\mathrm{cor}^r_{S,G}.$$

Similarly the right square is commutative. It follows from 8.5 that the diagram

$$\begin{array}{ccccccccc}
\cdots \to & H^r(S, M) & \to & H^r(S, N) & \xrightarrow{\delta^*} & H^{r+1}(S, L) & \to & H^{r+1}(S, M) & \to \cdots \\
& \downarrow \mathrm{cor}^r_{S,G} & & \downarrow \mathrm{cor}^r_{S,G} & & \downarrow \mathrm{cor}^{r+1}_{S,G} & & \downarrow \mathrm{cor}^{r+1}_{S,G} & \\
\cdots \to & H^r(G, M) & \to & H^r(G, N) & \xrightarrow{\delta^*} & H^{r+1}(G, L) & \to & H^{r+1}(G, M) & \to \cdots
\end{array}$$

is commutative.

32.5 Let M, N be G-modules, $S \leqq G$, and $f\colon M \to N$ an S-homomorphism. We consider here the homomorphism

$$\mathrm{cor}^r_{S,G} \cdot f^* \cdot \mathrm{res}^r_{G,S}$$

where $f^*\colon H^r(S, M) \to H^r(S, N)$ is the homomorphism induced by f. Let X_* be a complete resolution of G (so, also of S). Let $t \in \mathrm{Hom}_G(X_r, M)$ be a cocycle representing $\bar{t} \in H^r(G, M)$. Then t viewed as an S-homomorphism is a cocycle representing $\mathrm{res}^r_{G,S}\bar{t}$. Thus ft represents $f^* \cdot \mathrm{res}^r_{G,S}\bar{t} \in H^r(S, N)$, and $T^r_{S,G}(ft)$ represents $\mathrm{cor}^r_{S,G} \cdot f^* \cdot \mathrm{res}_{G,S}\bar{t} \in H^r(G, N)$. Since t and $T_{S,G}(f)$ are G-homomorphisms, we conclude that

$$\mathrm{cor}^r_{S,G} \cdot f^* \cdot \mathrm{res}^r_{G,S} = (T_{S,G}(f))^r\colon H^r(G,M) \to H^r(G, N) \tag{9}$$

where $(T_{S,G}(f))^r$ is the $\mathbf{Z}$-homomorphism induced by the G-homomorphism $T_{S,G}(f)\colon \mathrm{Hom}_G(X_r, M) \to \mathrm{Hom}_G(X_r, N)$.

32.6 An interesting special case of (9) is obtained when $f = 1_M$. In this case

$$\mathrm{cor}^r_{S,G} \cdot 1^*_M \cdot \mathrm{res}^r_{G,S} = \left(\sum g_i 1_M g_i^{-1}\right)^* = (G\colon S)^*$$

where $(G\colon S)$ is the index of S in G. Thus:

$$\mathrm{cor}^r_{S,G}\ \mathrm{res}^r_{G,S}\colon H^r(G, M) \to H^r(G, M)$$

is multiplication by the index of S in G.

33. Sylow Subgroups

33.1 Let G be a group of order n. For any prime p, write

$$n = p^{v_p} \cdot m_p$$

such that p and m_p are relatively prime. The subgroups of order p^{v_p} are called p-Sylow subgroups of G. Let M be a G-module, and

$$H^r(G, M)_p = \{x \in H^r(G, M), \text{ order of } x \text{ is a power of } p\}.$$

$H^r(G, M)_p$ is called the *p-primary component of* $H^r(G, M)$. We have

$$H^r(G, M) = \sum_{p|n} H^r(G, M)_p.$$

Let $\mathrm{inj}_p \colon H^r(G, M)_p \to H^r(G, M)$ be the injection of the pth summand into the direct sum, and $\mathrm{proj}_p \colon H^r(G, M) \to H^r(G, M)_p$ the projection onto the summand $H^r(G, M)_p$. Then *if G_p is a p-Sylow subgroup of G,*

(i) $$0 \to H^r(G, M)_p \xrightarrow{\mathrm{res}^r_{G,G_p} \cdot \mathrm{inj}_p} H^r(G_p, M)$$

is exact,

(ii) $$H^r(G_p, M) \xrightarrow{\mathrm{proj}_p \cdot \mathrm{cor}^r_{G_p,G}} H^r(G, M)_p \to 0$$

is exact,

(iii) $$H^r(G_p, M) \cong \mathrm{im}(\mathrm{res}^r_{G,G_p}) \oplus \ker(\mathrm{cor}^r_{G_p,G}).$$

Proof Let $n = m_p \cdot p^{v_p}$ be the order of G where $(m_p, p) = 1$. Then there exist integers m'_p, s such that

$$m_p m'_p + s p^{v_p} = 1. \tag{10}$$

Let $\bar{t} \in H^r(G, M)'_p$, then

$$\bar{t} = m_p m'_p \bar{t} = \mathrm{cor}^r_{G_p,G}\, m'_p\, \mathrm{res}^r_{G,G_p} \bar{t}.$$

Thus res_{G,G_p} is a monomorphism when restricted to $H^r(G, M)_p$, and $\mathrm{cor}^r_{G_p,G}$ is an epimorphism when followed by proj_p. This establishes (i) and (ii). To show (iii), for $\beta \in H^r(G_p, M)$ write

$$\beta = m_p' \operatorname{res}^r_{G,G_p} \operatorname{cor}^r_{G_p,G} \beta + (\beta - m_p' \operatorname{res}^r_{G,G_p} \operatorname{cor}^r_{G_p,G} \beta). \tag{11}$$

The first term on the right is in $\operatorname{im}(\operatorname{res}^r_{G,G_p})$. Applying $\operatorname{cor}^r_{G_p,G}$ to the second term on the right we get $s \cdot p^{v_p} \operatorname{cor}^r_{G_p,G} \beta = 0$ by (10) and (ii). Moreover if $(q, p) = 1$, $\operatorname{res}^r_{G,G_p}$ maps $H^r(G, M)_q$ to zero by 32.6. and 15.5. Hence

$$H^r(G_p, M) \cong \operatorname{im}(\operatorname{res}^r_{G,G_p}) + \ker(\operatorname{cor}^r_{G_p,G}).$$

To show this is a direct sum suppose $\beta = \operatorname{res}^r_{G,G_p} \alpha$ and $\operatorname{cor}^r_{G_p,G} \beta = 0$. Then, by 32.6, $\operatorname{cor}^r_{G_p,G} \operatorname{res}^r_{G,G_p} \alpha = m_p \alpha = 0$. Since $\alpha \in H^r(G, M)_p$ and $(m_p, p) = 1$ we have $\alpha = 0$; hence $\beta = 0$.

We conclude:

(iv) *If G is of order n and G_p is a p-Sylow subgroup, then the map*

$$\sum_{p|n} \operatorname{res}^r_{G,G_p} : H^r(G, M) = \sum_{p|n} H^r(G, M)_p \to \sum_{p|n} H^r(G_p, M)$$

is a monomorphism, and moreover the image is a direct summand.

(v) *If $H^r(G_p, M)$ is finite for every $p|n$, then $H^r(G, M)$ is finite.*

(vi) *If $H^r(G_p, M) = 0$ for every $p|n$, then $H^r(G, M) = 0$.*

34. Generalization of G-Induced Modules

34.1 *Let $S \leqq G$, $G = \dot{\bigcup} Sg$, where we choose $g = 1$ for the coset S. Suppose M is a G-module and $N \subset M$ is an S-submodule such that*

$$M = \sum gN,$$

then,

$$H^r(G, M) \cong H^r(S, N). \tag{12}$$

Remark If $S = \{1\}$, then N above is a $\mathbf{Z}$-module and thus M is G-induced, and so $H^r(G, M) = 0$ (see 15.4). This is a special case of (12).

Proof of (12) Let $\operatorname{proj}_1 : M \to N$ be the projection of M onto the summand $1 \cdot N$. If $m \in M$, $m = \sum gn_g$, $n_g \in N$, then $\text{proj}_1\, m = n_1$. For $s \in S$, $\text{proj}_1\, sm = sn_1 = s\, \text{proj}_1\, m$. Thus proj_1 is an S-homomorphism. Let $i: N \to M$ be the S-injection. Then $i \cdot \text{proj}_1: M \to M$ is an S-homomorphism. By 32.5, (9) we have

$$\text{cor}^r_{S,G}(i \cdot \text{proj}_1)^* \text{res}^r_{G,S} = T_{S,G}(i \cdot \text{proj}_1)^r$$

and

$$T_{S,G}(i \cdot \text{proj}_1)(\sum_{g \in G} gn_g) = \sum_{h \in G} h(i \cdot \text{proj}_1) \sum_{g \in G} h^{-1} gn_g = \sum_{h \in G} hn_h.$$

Therefore $T_{S,G}(i \cdot \text{proj}_1) = 1_M$, and so

$$\text{cor}^r_{S,G} \cdot i^* \cdot \text{proj}^*_1\, \text{res}^r_{G,S} = 1_{H^r(G,M)}.$$

Next, if $t \in \text{Hom}_S(X_r, N)$ is a cocycle, then $it \in \text{Hom}_S(X_r, M)$, $T_{S,G}(it) \in \text{Hom}_G(X_r, M)$, and $\text{proj}_1\, T_{S,G}(it) \in \text{Hom}_S(X_r, N)$. Since

$$\text{proj}_1\, T_{S,G}(it) = \text{proj}_1(\sum gitg^{-1}) = it,$$

we have

$$\text{proj}^*_1\, \text{res}^r_{G,S}\, \text{cor}^r_{S,G} i^* = 1_{H^r(S,N)}.$$

Thus

$$\text{cor}^r_{S,G} i^*: H^r(S, N) \to H^r(G, M)$$

and

$$\text{proj}^*_1\, \text{res}^r_{G,S}: H^r(G, M) \to H^r(S, N)$$

are inverses of each other.

34.2 (*Shapiro's lemma*) *Let $S \leqq G$ and M an S-module. Then $\mathbf{Z}G \otimes_{\mathbf{Z}S} M$ is a (left) G-module and*

$$H^r(G, \mathbf{Z}G \otimes_{\mathbf{Z}S} M) \cong H^r(S, M).$$

Proof Write $G = \bigcup_{i=1}^n g_i S$ as a disjoint union of left cosets of S. Then every element of $\mathbf{Z}G \otimes_{\mathbf{Z}S} M$ can be written uniquely in the form $\sum_{i=1}^n g_i \otimes m_i$. Hence

$$\mathbf{Z}G \otimes_{\mathbf{Z}S} M = \sum_{i=1}^n g_i \otimes M \qquad \text{(direct)}.$$

Our assertion now follows from 34.1.

REMARK Under the hypothesis of 34.2 one can similarly show that

$$H^r(G, \mathrm{Hom}_{\mathbf{Z}S}(\mathbf{Z}G, M)) \cong H^r(S, M)$$

for all $r \in \mathbf{Z}$. This follows from 34.1 by observing that $\mathrm{Hom}_{\mathbf{Z}S}(\mathbf{Z}G, M) = \sum_{i=1}^{n} M_i$ (direct) with all $M_i = M$.

35. The Inflation Map

In this section we study the properties of normal subgroups of G. The results obtained here are valid only for cohomology groups of positive dimension.

35.1 Let $S \trianglelefteq G$, and M be a G-module. Then $M^S = \{m \in M, sm = m \ \forall\, s \in S\}$ is a G/S-module by $(g_i S)m = g_i m$. This action of G/S on M^S is well defined, for,

$$(g_i sS)m = g_i sm = g_i m \qquad \text{for every} \quad s \in S, \quad m \in M^S.$$

To verify that $(g_i S)m \in M^S$, we observe that $s(g_i S)m = sg_i m = g_i s' m = g_i m$. Let $i\colon M^S \to M$ be the inclusion G-homomorphism and $\lambda\colon G \to G/S$ the canonical epimorphism. Then the map

$$(\lambda, i)_r^*\colon H^r(G/S, M^S) \to H^r(G, M)$$

of 31.3 is well defined for $r \geqq 1$; $(\lambda, i)_r^*$ is called the *inflation map* and is denoted by $\mathrm{inf}_{G/S,G}^r$ or simply by inf.

35.2 In 35.6 we will show that under suitable hypothesis the sequence

$$0 \to H^r(G/S, M^S) \xrightarrow{\text{inf}} H^r(G, M) \xrightarrow{\text{res}} H^r(S, M)$$

is exact. We show here a part of this assertion, namely that

$$\mathrm{res}_{G,S}^r \, \mathrm{inf}_{G/S,G}^r = 0. \tag{13}$$

Let $\lambda\colon G \to G/S$ be the canonical epimorphism. Consider the commutative diagram

$$
(14)\qquad
\begin{array}{ccccccccc}
0 & \leftarrow & \mathbf{Z} & \leftarrow & X_0 & \leftarrow & X_1 & \leftarrow & \cdots \\
 & & \Big\downarrow 1_{\mathbf{Z}} & & \Big\downarrow \Lambda_0 & & \Big\downarrow \Lambda_1 & & \\
0 & \leftarrow & \mathbf{Z} & \leftarrow & \bar{X}_0 & \leftarrow & \bar{X}_1 & \leftarrow & \cdots
\end{array}
$$

where the upper (resp. lower) row is the normalized bar resolution of the trivial G-module (resp. G/S-module) $\mathbf{Z}$, and

$$\Lambda_r g[g_1, \ldots, g_r] = \lambda g[\lambda g_1, \ldots, \lambda g_r] = gS[g_1 S, \ldots, g_r S].$$

Then the diagram

$$
\begin{array}{ccc}
\operatorname{Hom}_{G/S}(\bar{X}^r, M^S) & \xrightarrow{(\Lambda_r,\, i)'} & \operatorname{Hom}_G(X_r, M) \\
\Big\downarrow (1_{X_r}, 1_{M^S})' & & \Big\downarrow (1_{X_r}, 1_M)' \\
\operatorname{Hom}_{S/S}(\bar{X}_r, M^S) & \xrightarrow{(\Lambda_r,\, i)'} & \operatorname{Hom}_S(X_r, M)
\end{array}
$$

is commutative. Therefore the corresponding diagram of the induced homomorphisms

$$
\begin{array}{ccc}
H^r(G/S, M^S) & \xrightarrow{\inf^r_{G/S,G}} & H^r(G, M) \\
\Big\downarrow \operatorname{res}^r_{G/S,S/S} & & \Big\downarrow \operatorname{res}^r_{G,S} \\
H^r(S/S, M^S) & \xrightarrow{\inf^r_{S/S,S}} & H^r(S, M)
\end{array}
$$

is commutative. The identity (13) follows from the fact that $H^r(S/S, M^S) = 0$.

35.3 *The sequence*

$$0 \to H^1(G/S, M^S) \xrightarrow{\inf^1_{G/S,G}} H^1(G, M) \xrightarrow{\operatorname{res}^1_{G,S}} H^1(S, M)$$

is exact.

PROOF We will use diagram (14) of 35.2 in this proof. Let $\bar{t} \in H^1(G/S, M^S)$ such that $\inf_{G/S,G} \bar{t} = 0$. Let $t \in \operatorname{Hom}_{G/S}(\bar{X}_1, M^S)$ be a normalized cocycle representing $\bar{t}$. Then $(\Lambda_1, i)'t \in \operatorname{Hom}_G(X_1, M)$ is a principal homomorphism (see 19.1), i.e.,

$$((\Lambda_1, i)'t)[g] = it[gS] = gm - m$$

for some $m \in M$. We claim $m \in M^S$, for, if $s \in S$, $sm - m = it[sS] = 0$. Hence if $g \in G$,

$$it[gS] = gm - m = i(gSm - Sm).$$

Since i is a monomorphism,

$$t[gS] = gSm - Sm,$$

so t is a coboundary. To show exactness at $H^1(G, M)$, let $\bar{t} \in H^1(G, M)$ such that $\operatorname{res}^1_{G,S} \bar{t} = 0$. Let $t \in \operatorname{Hom}_G(X_1, M)$ be a normalized cocycle representing $\bar{t}$. Then $t[s] = sm - m$, for $s \in S$, $m \in M$. The cocycle $t' \in \operatorname{Hom}_G(X_1, M)$ defined by $t'[g] = t[g] - (gm - m)$ represents $\bar{t}$ also but has the advantage that $t'[s] = 0$ for $s \in S$. Since t' is a normalized cocycle,

$$(\delta t')[g, h] = gt'[h] - t'[gh] + t'[g] = 0. \tag{15}$$

I. Choose $h \in S$ in (15), then

$$t'[gh] = t'[g].$$

This means t' is a step function (value of t' is constant on any representative of a coset gS).

II. Choose $g \in S$ in (15), then

$$\begin{aligned} gt'[h] = t'[gh] &= t'[hg'] \qquad \text{for some} \quad g' \in S, \qquad \text{since} \quad S \trianglelefteq G \\ &= t'[h] \qquad \text{by I.} \end{aligned}$$

This means $t'[h] \in M^S$ for any $h \in G$. Thus t' induces a cochain $\bar{t}' \in \operatorname{Hom}_{G/S}(\bar{X}_1, M^S)$ defined by

$$\bar{t}'[gS] = t'[g].$$

Since $t'[g] = t[g]$ modulo coboundaries, we have established exactness at $H^1(G, M)$.

35.4 *Let*

$$0 \to L \xrightarrow{i} M \xrightarrow{j} N \to 0$$

be an exact sequence of G-homomorphisms. If $H^1(G, L) = 0$, *then*

$$0 \to L^G \xrightarrow{i|_{L^G}} M^G \xrightarrow{j|_{M^G}} N^G \to 0$$

is exact.

PROOF The verification that the sequence is left exact is trivial. We will show $j|_{M^G}$ is onto. Let $n \in N^G$, then $n = jm$ for some $m \in M$. This implies $gm - m = il_g$ for some $l_g \in L$ which depends on g. Since $l_1 = 0$, l_g is a normalized 1-cochain. Moreover, since

$$i(gl_h - l_{gh} + l_g) = g(hm - m) - (ghm - m) + gm - m = 0,$$

l_g is a 1-cocycle. Since $H^1(G, L) = 0$, there exists $\lambda \in L$ such that $l_g = g\lambda - \lambda$ (i.e., l_g is a coboundary), and so $i(g\lambda - \lambda) = il_g = gm - m$. This implies $g(m - i\lambda) - (m - i\lambda) = 0$, so $m - i\lambda \in M^G$ and $j(m - i\lambda) = n$.

35.5 *If M is G-induced, then* M^S *is G/S-induced for any* $S \trianglelefteq G$.

PROOF Let $M = \sum_{g\in G} gN$ (direct). Let $m \in M^S \subseteq M$, then $m = \sum_{g\in G} gn_g$. Since $s^{-1}m = m$ for $s \in S$, we have

$$\sum_{g\in G} gn_g = \sum_{g\in G} s^{-1}gn_g = \sum_{g\in G} gn_{sg};$$

so $s^{-1}m = m$ if and only if $n_g = n_{sg}$. Therefore $m = \sum_{i=1}^{r} g_i T_{\{1\},S}(n_{g_i})$ uniquely, where $G = \dot{\bigcup}_{i=1}^{r} g_i S$. Therefore $M^S = \sum_{i=1}^{r} g_i T_{\{1\},S}(N)$ where $T_{\{1\},S}(N)$ is the **Z**-module generated by the traces of elements of N.

35.6 *Let S be a normal subgroup of G. Assume*

(16) $$H^i(S, M) = 0 \qquad \text{for} \quad i = 1, 2, \ldots, r - 1,$$

then

$$(17) \qquad 0 \to H^r(G/S, M^S) \xrightarrow{\text{inf}} H^r(G, M) \xrightarrow{\text{res}} H^r(S, M)$$

is exact.

REMARK Under the hypothesis (16) the sequence (17) can be replaced by an exact sequence

$$0 \to H^r(G/S, M^S) \to H^r(G, M) \to H^r(S, M)^{G/S} \to H^{r+1}(G/S, M^S) \to H^{r+1}(G, M).$$

The exactness of this sequence is established using spectral sequences in Chapter 8.

PROOF OF (17) Induction on r. The sequence is exact for $r = 1$ by 35.3. Let $r > 1$ and suppose the assertion in 35.6 is true for dimensions less than r. Then by 35.4 and 16.1,(2) we can construct the commutative diagram

$$(18) \qquad \begin{array}{ccccccccc} 0 \to & M & \to & P & \to & N & \to 0 \\ & \uparrow & & \uparrow & & \uparrow & \\ 0 \to & M^S & \to & P^S & \to & N^S & \to 0 \end{array}$$

where the vertical maps are inclusions and P is G-induced [see remarks following 16.1,(2)]. Hence by 35.5, P^S is G/S-induced. Thus for every i we have

$$0 = H^{i-1}(G, P) \to H^{i-1}(G, N) \overset{\delta^*}{\cong} H^i(G, M) \to H^i(G, P) = 0.$$

Similarly

$$H^{i-1}(S, N) \cong H^i(S, M)$$

and

$$H^{i-1}(G/S, N^S) \cong H^i(G/S, M^S).$$

By diagram (18) and 31.4, we have the commutative diagram

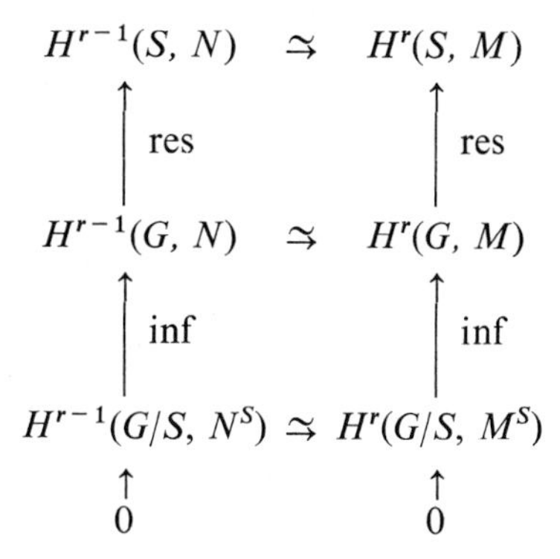

where the left column is exact by the induction hypothesis. Hence the right column is exact.

36. Cohomological Triviality

Let G be a finite group. The G-module M is *cohomologically trivial* if for every subgroup S of G and for every integer r we have $H^r(S, M) = 0$.

36.1 (*The Nakayama–Tate criterion.*) *Let G be a finite group and M a G-module. Then $H^i(S, M) = 0$ for two successive i and all subgroups S of G if and only if M is cohomologically trivial.*

REMARK Nakayama's criterion for cohomological triviality is in [76] and [77]. For other criteria concerning cohomological triviality see D. S. Rim [90] L. Evens [30] and Onishi [83].

We will prove some preliminary results in 36.2–36.4, then using these and dimension shifters we will prove in 36.5 the Nakayama–Tate criterion for cohomological triviality.

36.2 *Let $S \trianglelefteq G$ and M a G-module. Then there exists an exact sequence*

(19) $$0 \leftarrow H^0(G/S, M^S) \leftarrow H^0(G, M) \leftarrow H^0(S, M).$$

PROOF By 17.2, it is enough to show the existence of an exact sequence

$$0 \leftarrow M^G/T_{S,G}(M^S) \overset{\varphi}{\leftarrow} M^G/T_G(M) \overset{\psi}{\leftarrow} M^S/T_S(M). \tag{20}$$

Define φ by $\varphi(m + T_G(M)) = m + T_{S,G}(M^S)$ for $m \in M^G$; φ is well defined and onto since

$$T_G(M) = T_{S,G}T_S(M) \subset T_{S,G}(M^S).$$

Define ψ by $\psi(m + T_S(M)) = T_{S,G}(m) + T_G(M)$; ψ is well defined for if $m + T_S(m') + T_S(M)$ is another representation of the coset $m + T_S(M)$, then

$$\begin{aligned}\psi(m + T_S(m') + T_S(M)) &= T_{S,G}(m) + T_{S,G}T_S(m') + T_G(M)\\ &= T_{S,G}(m) + T_G(m') + T_G(M)\\ &= T_{S,G}(m) + T_G(M).\end{aligned}$$

The sequence (20) is a complex, for,

$$\varphi\psi(m + T_S(M)) = T_{S,G}(m) + T_{S,G}(M^S) = T_{S,G}(M^S).$$

Furthermore, (20) is exact at $M^G/T_G(M)$, for, if $\varphi(m + T_G(M)) = T_{S,G}(M^S)$, then $m = T_{S,G}(m')$ for some $m' \in M^S$. We observe that $m' + T_S(M) \in M^S/T_S(M)$, and moreover

$$\psi(m' + T_S(M)) = T_{S,G}(m') + T^G(M) = m + T_G(M).$$

Thus (20) is exact.

36.3 *Let G be a p-group for some prime p, then:*

(i) *If $H^0(S, M) = H^1(S, M) = 0$ for all $S \leqq G$, then $H^2(S, M) = 0$ for all $S \leqq G$.*

(ii) *If $H^1(S, M) = H^2(S, M) = 0$ for all $S \leqq G$, then $H^0(S, M) = 0$ for all $S \leqq G$.*

PROOF There exists $N \lhd G$ such that G/N is prime cyclic. Consider the diagram

$$
(21)\qquad \begin{array}{c}
0 \leftarrow H^0(G/N, M^N) \leftarrow H^0(G, M) \leftarrow H^0(N, M) \\
\Big\downarrow \wr \\
0 \rightarrow H^2(G/N, M^N) \xrightarrow{\text{inf}} H^2(G, M) \xrightarrow{\text{res}} H^2(N, M)
\end{array}
$$

where the upper row is the exact sequence of 36.2, the lower row is exact since $H^1(N, M) = 0$ (35.6), and the vertical map is an isomorphism by periodicity of the cohomology groups of the cyclic group G/N (Section 23). We now proceed by induction on the order of G. If $|G| = 1$, (i) and (ii) are satisfied. Suppose (i) and (ii) true for all groups of order less than $|G|$. To show (i), $H^0(N, M) = H^2(N, M) = 0$ by the induction hypothesis. Then by (21)

$$H^0(G, M) \cong H^0(G/N, M^N) \cong H^2(G/N, M^N) \cong H^2(G, M).$$

Since by the hypothesis of (i), $H^0(G, M) = 0$, we conclude that $H^2(G, M) = 0$. To show (ii), $H^0(N, M) = H^2(N, M) = 0$ by the induction hypothesis. Since in this case $H^2(G, M) = 0$, by (21)

$$0 = H^2(G, M) \cong H^2(G/N, M^N) \cong H^0(G/N, M^N) \cong H^0(G, M).$$

36.4 *The statement of* **36.3** *holds if G is any finite group.*

Proof If G is not a prime power, for each prime $p \| G|$ let G_p be a p-Sylow subgroup. Then $|G_p| < |G|$ and, by 36.3, it is enough to show $H^0(G_p, M) = H^2(G_p, M) = 0$ implies $H^0(G, M) = H^2(G, M) = 0$. If $H^i(G, M)_p$, $i = 0, 2$, is the p-primary part of $H^i(G, M)$, $i = 0, 2$, the sequence

$$H^i(G_p, M) \xrightarrow{\text{proj}_p \cdot \text{cor}_{G_p, G}} H^i(G, M)_p \rightarrow 0$$

is exact [33.1,(ii)]. Therefore $H^i(G, M)_p = 0$, $i = 0, 2$. Hence $H^0(G, M) = H^2(G, M) = 0$.

36.5 *Proof of the Nakayama–Tate criterion* (**36.1**). Suppose $H^r(S, M) = H^{r+1}(S, M) = 0$ for all $S \leqq G$. By repeated use of dimension shifters there exists a G-module N such that

$$(22)\qquad H^n(S, N) \cong H^{n+r}(S, M)$$

for any $n \in \mathbf{Z}$ and any subgroup $S \leqq G$. So, in particular, for $n = 0, 1$, we have

$$H^0(S, N) \cong H^r(S, M) = 0 \qquad \text{and} \qquad H^1(S, N) \cong H^{r+1}(S, M) = 0.$$

Thus, by part (i) of 36.4, $H^2(S, N) = 0$ for all $S \leqq G$. This implies, by (22),

$$H^{r+2}(S, M) = 0 \qquad \text{for all} \quad S \leqq G.$$

Repeating this argument and using $H^{r+1}(S, M) \cong H^{r+2}(S, M) = 0$ we get $H^{r+3}(S, M) = 0$. Hence, by induction, $H^t(S, M) = 0$ for all $S \leqq G$ if $t \geqq r$. A similar argument using part (ii) of 36.4 shows $H^t(S, M) = 0$ for all $S \leqq G$ and $t \leqq r + 1$.

37. Cohomological Equivalence

Let $f: M \to N$ be a G-homomorphism. If for all $S \leqq G$ and all $r \in \mathbf{Z}$, $f^*: H^r(S, M) \to H^r(S, N)$ is an isomorphism, then f is a *cohomological equivalence*. We will also say M and N are *cohomologically equivalent under* f.

37.1 *Let $f: M \to N$ be a G-homomorphism. Then there exists a G-module M'' and homomorphisms δ^*, g^* such that for all $S \leqq G$ the sequence*

$$\cdots \xrightarrow{\delta^*} H^{r-1}(S, M) \xrightarrow{f^*} H^{r-1}(S, N) \xrightarrow{g^*} H^{r-1}(S, M'')$$
$$\xrightarrow{\delta^*} H^r(S, M) \xrightarrow{f^*} \cdots$$

is exact.

Proof Let M' be G-induced such that there exists a monomorphism $i: M \to M'$ [see 17.1,(2)]. Let $(f, i): M \to N \oplus M'$ be the G-homomorphism such that $\text{proj}_N(f, i) = f$ and $\text{proj}_{M'}(f, i) = i$. Since i is a monomorphism, so is (f, i). Let $M'' = N \oplus M'/(f, i)M$. Then in the diagram

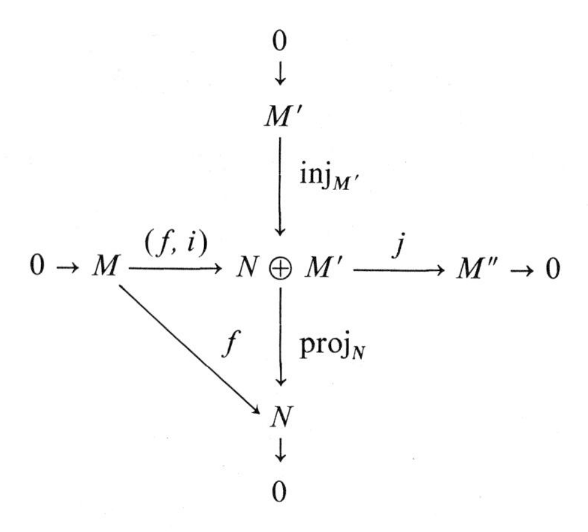

the row and the column are exact, where j is the canonical epimorphism. Since M' is G-induced, we have

$$\begin{array}{ccccccc}
 & & & H^r(S, M') = 0 & & & \\
 & & & \downarrow & & & \\
\cdots \xrightarrow{\delta^*} & H^r(S, M) & \xrightarrow{(f,\, i)^*} & H^r(S, N \oplus M') & \xrightarrow{j^*} & H^r(S, M'') \to \cdots & \\
 & & \searrow^{f^*} & \downarrow \mathrm{proj}_N^* & & & \\
 & & & H^r(S, N) & & & \\
 & & & \downarrow \delta^* & & & \\
 & & & H^{r+1}(S, M') = 0. & & &
\end{array}$$

Let $g^* = j^*(\mathrm{proj}_N^*)^{-1}$. Then we have the desired exact sequence.

37.2 *(Tate's criterion.) Let $f\colon M \to N$ be a G-homomorphism. Then f is a cohomological equivalence if and only if for every $S \leqq G$,*

(i) $f_{r-1}^*\colon H^{r-1}(S, M) \to H^{r-1}(S, N)$ *is an epimorphism,*

(ii) $f_r^*\colon H^r(S, M) \to H^r(S, N)$ *is an isomorphism,*

(iii) $f_{r+1}^*\colon H^{r+1}(S, M) \to H^{r+1}(S, N)$ *is a monomorphism.*

Proof Conditions (i)–(iii) obviously hold if f is a cohomological equivalence. Conversely, consider the exact sequence of 37.1. Condition (i) implies $g^*_{r-1} = 0$ and (ii) implies $\delta^*_{r-1} = 0$. Thus $H^{r-1}(S, M'') = 0$. Similarly (ii) implies $g^*_r = 0$ and (iii) implies $\delta^*_r = 0$. Thus $H^r(S, M'') = 0$. The vanishing of $H^{r-1}(S, M'')$ and $H^r(S, M'')$ for all $S \leqq G$ implies, by 36.1, that $H^t(S, M'') = 0$ for all $t \in \mathbf{Z}$ and all $S \leqq G$. Thus f^*_t is an isomorphism for all t and all S.

Remark Evens [30] has shown that $f\colon M \to N$ *is a cohomological equivalence if and only if* f^*_{r-1}, f^*_r *are isomorphisms.*

Tate's and Evens' criteria for cohomolegical equivalence are special cases of Onishi's (Math. Z. 120 (1971) 221–223)) which states that: $f\colon M \to N$ *is a cohomological equuivalence if and only if there are integers* k, q, k', q' *with* $q - k$, $q' - k'$, $k - k'$ *odd,* $q - k \leqq 1$, $q' - k' \leqq 1$, *such that* $f^*_k, f^*_{k'}$ *are epimorphisms and* $f^*_q, f^*_{q'}$ *are monomorphisms.* This gives Tate's criterion for $q = k + 1 = k' = q' - 1$, and Evens' criterion for $q = k + 1 = k' = q' + 1$.

37.3 *Let M be a G-module. Suppose*

(i) $H^{-1}(S, M) = 0$ *for all* $S \leqq G$,

(ii) $H^0(S, M)$ *is cyclic of order equal to order of S for all* $S \leqq G$.

Then there exists a homomorphism $f\colon \mathbf{Z} \to M$ *such that f is a cohomological equivalence.*

Proof Let $\gamma = |G|$, $\sigma = |S|$, $\gamma = \delta\sigma$. Let ξ be a generator of $H^0(G, M)$. Then $\mathrm{cor}_{S,G}\,\mathrm{res}_{G,S}\,\xi = \delta\xi$ by 32.6. Hence $\mathrm{cor}_{S,G}\,\mathrm{res}_{G,S}\,\xi$ has order σ. Therefore $\mathrm{res}_{G,S}\,\xi$ has order at least σ. But since $H^0(S, M)$ has order σ, $\mathrm{res}_{G,S}\,\xi$ generates $H^0(S, M)$. Thus $\mathrm{res}_{G,S}$ maps a generator of $H^0(G, M)$ to a generator of $H^0(S, M)$ for any $S \leqq G$. Let

$$\kappa^M_{0,S}\colon M^S \to M^S/T_S(M) \overset{\varphi}{\cong} H^0(S, M)$$

be the canonical epimorphism followed by φ, then in the commutative diagram

$$\begin{array}{ccc} H^0(G, M) & \xrightarrow{\text{res}_{G,S}} & H^0(S, M) \\ \uparrow \kappa^M_{0,G} & & \uparrow \kappa^M_{0,S} \\ M^G & \xrightarrow{\text{inclusion}} & M^S \end{array}$$

there exists $m \in M^G$ such that $\kappa^M_{0,G}(m) = \xi$. Moreover the images of m under the $\kappa^M_{0,S}$ map m to the generators of the various cohomology groups $H^0(S, M)$. View $\mathbf{Z}$ as a trivial G-module and let $f: \mathbf{Z} \to M$ be defined by $f(1) = m$; f is a G-homomorphism; moreover, it is a cohomological equivalence. For:

(i) $f^*_{-1}: H^{-1}(S, \mathbf{Z}) \to H^{-1}(S, M)$

is an epimorphism since $H^{-1}(S, M) = 0$.

(ii) In the commutative diagram

$$\begin{array}{ccc} \mathbf{Z} = \mathbf{Z}^S & \xrightarrow{f} & M^S \\ \downarrow \kappa^{\mathbf{Z}}_{0,S} & & \downarrow \kappa^M_{0,S} \\ \mathbf{Z}/\sigma\mathbf{Z} = H^0(S, \mathbf{Z}) & \xrightarrow{f^*_0} & H^0(S, M) \end{array}$$

both $H^0(S, \mathbf{Z})$ and $H^0(S, M)$ are cyclic of order σ. Moreover $\kappa^{\mathbf{Z}}_{0,S}(1)$ is a generator of $H^0(S, \mathbf{Z})$, and

$$f^*_0 \kappa^{\mathbf{Z}}_{0,S}(1) = \kappa^M_{0,S} f(1) = \kappa^M_{0,S} m = \xi_S$$

is a generator of $H^0(S, M)$. Thus f^*_0 carries generator to generator. Since both $H^0(S, \mathbf{Z})$ and $H^0(S, M)$ are of order σ; f^*_0 is an isomorphism.

(iii) $H^1(S, \mathbf{Z}) = \text{Hom}(S, \mathbf{Z}) = 0$ because the action of S on $\mathbf{Z}$ is trivial and the only homomorphism from a finite group to $\mathbf{Z}$ is the trivial one. Therefore

$$f^*_1: H^1(S, \mathbf{Z}) \to H^1(S, M)$$

is a monomorphism.

37.4 (Tate [106]) *Let M be a G-module such that for all $S \leqq G$, $H^1(S, M) = 0$ and $H^2(S, M)$ is cyclic of order $|S|$. Then for all r and all $S \leqq G$*

$$H^r(S, M) \cong H^{r-2}(S, \mathbf{Z}) \tag{23}$$

PROOF Construct an exact sequence

$$0 \to M \to M' \to N \to 0$$

where M' is G-induced. Then $H^{r-1}(S, N) \cong H^r(S, M)$ for all $S \leqq G$. Repeat by constructing an exact sequence

$$0 \to N \to N' \to N'' \to 0$$

where N' is G-induced. Then $H^{r-2}(S, N'') \cong H^r(S, M)$ for all $S \leqq G$. Thus, in particular, $H^{-1}(S, N'') = 0$, and $H^0(S, N'') \cong H^2(S, M)$ is cyclic of order $|S|$ for all $S \leqq G$. By 37.3,

$$H^{r-2}(S, \mathbf{Z}) \cong H^{r-2}(S, N'') \cong H^r(S, M).$$

REMARK For a generalization of the isomorphism (23) see Nakayama [78] and Serre [93, Theorem 14, p. 156].

CHAPTER

6

Finite p-Nilpotent Groups

38. Outer Automorphisms of Finite p-Groups

38.1 (Hoechsmann [52]) *Let G be a finite p-group and M a G-module of p-power order such that $H^0(G, M) = 0$. Then $H^0(S, M) = 0$ for every $S \leqq G$.*

PROOF Since every subgroup of G is subnormal (i.e., is a member of some normal series of G), it is enough to show $H^0(S, M) = 0$ for maximal normal subgroups S in G. By the exactness of the sequence

(1) $$0 \leftarrow H^0(G/S, M^S) \leftarrow H^0(G, M) \leftarrow H^0(S, M)$$

given in 36.2, we have $(M^S)^{G/S} = M^G = T_{G/S}(M^S)$. Moreover, since M^S is finite, G/S is cyclic, and $H^0(G/S, M^S) = 0$, by 23.2, $H^{-1}(G/S, M^S) = 0$. Hence $(M^S)_{T_{G/S}} = (\bar{g} - 1)M^S$ where $\bar{g} = gS$ is a generator of G/S. Therefore the exact sequence

$$0 \rightarrow (M^S)_{T_{G/S}} \rightarrow M^S \rightarrow T_{G/S}M^S \rightarrow 0$$

turns upon the exact sequence

$$0 \rightarrow (\bar{g} - 1)M^S \rightarrow M^S \rightarrow T_{G/S}(T_S M) \rightarrow 0$$

since $M^G = T_G M = T_{G/S}(T_S M)$. Consider the commutative diagram

$$\begin{array}{ccccccccc} 0 \to & (\bar{g} - 1)M^S & \to & (\bar{g} - 1)M^S \oplus T_S M & \to & T_S M & \to 0 \\ & \downarrow \text{id.} & & \downarrow \rho & & \downarrow T_{G/S} & \\ 0 \to & (\bar{g} - 1)M^S & \longrightarrow & M^S & \xrightarrow{T_{G/S}} & T_{G/S}T_S M & \to 0 \end{array}$$

where the rows are exact and the map ρ is defined by the inclusions of $(\bar{g} - 1)M^S$ and $T_S M$ in M^S. By the Five Lemma, ρ is an epimorphism. Therefore

$$(\bar{g} - 1)M^S + T_S M = M^S. \tag{2}$$

Since $(\bar{g} - 1)^p M^S \subset pM^S$, substituting (2) for M^S in the first summand of (2) repeatedly, we obtain

$$pM^S + T_S M = M^S. \tag{3}$$

Similarly, substituting (3) for M^S in the left side of (3) repeatedly we obtain

$$T_S M = M^S.$$

Hence $H^0(S, M) = M^S/T_S M = 0$.

38.2 *Let G be a finite p-group, and M a G-module of p-power order. Then $H^n(G, M) = 0$ for $n \in \mathbf{Z}$ implies $H^n(S, M) = 0$ for all $S \leqq G$.*

PROOF This follows from 38.1 by dimension shifters (17.1).

38.3 *Let G be a finite p-group, and M a G-module of p-power order. Then $H^n(G, M) = 0$ if and only if $H^{n+1}(G, M) = 0$.*

PROOF *Case $n = 0$.* As in the proof of 38.1, using induction we may assume the statement true for maximal normal subgroups $S \leqq G$. Suppose $H^0(G, M) = 0$. Then, by the exactness of the sequence (1) of 38.1, $H^0(G/S, M^S) = 0$. Since G/S is cyclic and M^S is finite, by 23.2, $H^1(G/S, M^S) = 0$. Moreover by the induction hypothesis $H^1(S, M) = 0$. Hence by the exact inflation–restriction sequence

$$0 \to H^1(G/S, M^S) \xrightarrow{\text{inf}} H^1(G, M) \xrightarrow{\text{res}} H^1(S, M) \tag{4}$$

(35.3), $H^1(G, M) = 0$. Conversely, suppose $H^1(G, M) = 0$. Then by (4), $H^1(G/S, M^S) = 0$. Therefore, by 23.2, $H^0(G/S, M^S) = 0$. Since by the induction hypothesis $H^0(S, M) = 0$, it follows from the exactness of (1) that $H^0(G, M) = 0$.

Case $n \neq 0$. Follows from case $n = 0$ by dimension shifters.

38.4 (Gaschütz [34]) *Let G be a finite p-group, and M a G-module of p-power order. If $H^1(G, M) = 0$, then M is cohomologically trivial (i.e., $H^n(S, M) = 0$ for all n and all $S \leqq G$).*

PROOF Follows from 38.2 preceded by successive applications of 38.3.

38.5 *Let G be a p-group and M a maximal normal abelian subgroup. Then $M = C_G(M)$ where $C_G(M)$ is the centralizer of M in G.*

PROOF Suppose $M \lneqq C_G(M)$. Since $M \lhd G$, $C_G(M) \trianglelefteq G$, so $C_G(M)/M \trianglelefteq G/M$. Moreover G is a p-group, hence $C_G(M)/M \cap Z(G/M) \neq \{1\}$ where $Z(G/M)$ is the center of the group G/M. Let $g \in G$ be such that its canonical image $\bar{g} \in (C_G(M)/M \cap Z(G/M)$ is not 1. Then the subgroup of G generated by g and M is abelian normal and strictly contains M, contradicting the maximality of M.

38.6 (Gaschütz [34]) *If $p \neq 2$ and G is a finite p-group, then G has outer automorphisms.*

PROOF Let M (composition written additively) be a maximal normal abelian subgroup of G and let

$$0 \to M \xrightarrow{i} G \xrightarrow{j} \bar{G} \to 1$$

be an exact sequence. Then M has a canonical structure of $\bar{G}$-module given by conjugation in G (20.1). If $H^1(\bar{G}, M) = 0$, by 38.3, $H^2(\bar{G}, M) = 0$. Then, by 20.2, G is the semidirect product of M with $\bar{G}$, i.e., $G = M \times_0 \bar{G}$ where the subscript 0 denotes the zero factor set, so the composition in $M \times_0 \bar{G}$ is given by

$$(m_1, \bar{g}_1)(m_2, \bar{g}_2) = (m_1 + \bar{g}_1 m_2, \bar{g}_1\bar{g}_2).$$

Since the map $\alpha\colon M \times_0 \bar{G} \to M \times_0 \bar{G}$ defined by

$$(m, \bar{g}) = (-m, \bar{g})$$

is an automorphism of order 2, α is an outer automorphism.

If $H^1(\bar{G}, M) \neq 0$, let X_* be the standard normalized complete resolution of $\bar{G}$ and $\varphi\colon X_1 \to M$ a cocycle representing a nonzero element of $H^1(\bar{G}, M)$. Then if $\bar{g} = jg$ and $\bar{h} = jh$,

$$\varphi[\bar{g}\bar{h}] = \varphi[\bar{g}] + \bar{g}\varphi[\bar{h}]$$

and

$$\text{(5)} \qquad \varphi[\bar{g}] \neq m - \bar{g}m \qquad \text{for all} \quad m \in M$$

(i.e., φ is a crossed homomorphism but not a principal homomorphism). The map

$$\beta\colon G \to G$$

defined by

$$\text{(6)} \qquad \beta(g) = i\varphi[\bar{g}]\cdot g$$

is an automorphism and is such that $\beta \neq 1_G$ but $\beta|_M = 1_M$. For,

$$\begin{aligned}\beta(gh) &= i\varphi[\bar{g}\bar{h}]\cdot gh \\ &= i\varphi[\bar{g}]\bar{g}\cdot i\varphi[\bar{h}]\cdot gh \\ &= i\varphi[\bar{g}]\cdot gi\varphi[\bar{h}]g^{-1}gh \\ &= \beta(g)\cdot\beta(h);\end{aligned}$$

moreover, if $\beta(g) = 1$, then by (6), $g^{-1} = i\varphi(\bar{g}) \in iM$, and so $g = 1$ since $\beta|_M = 1_M$. Thus $\beta\colon G \to G$ is an automorphism since it is a monomorphism and G is finite. We have used $\beta|_M = 1_M$ and this holds since

$$\beta m = i\varphi[jm]\cdot m = i\varphi[1]\cdot m = m.$$

If β is an inner automorphism, then β is conjugation by an element of $C_G(M)$ since $\beta|_M = 1_M$. By 38.5, for some $m \in M$

$$\beta g = (im)g(im)^{-1}.$$

Therefore by (6)

$$i\varphi[\bar{g}] = img(i(-m))g^{-1} = im \cdot i\bar{g}(-m) = i(m - \bar{g}m).$$

Thus $\varphi[\bar{g}] = m - \bar{g}m$, which contradicts (5).

REMARK Gaschütz [35] has also shown that *if G is a nonabelian finite p-group, then G has an outer automorphism of p-power order*. A proof of this in detail is included in Gruenberg [45].

39. Cohomological Characterization of Finite p-Nilpotent Groups

The results in this section are due to Hoechsmann *et al.* [53].

Let p be a prime number. A finite group G is *p-nilpotent* if G contains a normal Hall subgroup K of p-power index, i.e., $(|K|, p) = 1$ and index of K is a power of p. This subgroup K is uniquely determined by p and is called the p-kernel of G. It can be shown that G is isomorphic to the semidirect product $K \cdot (G/K)$ (see [56]). A finite group G is *nilpotent* if and only if it is p-nilpotent for every prime number p (see [56]).

39.1 *Let G be a finite p-nilpotent group and M a G-module of p-power order. Then $H^n(G, M) = 0$ if and only if $H^{n+1}(G, M) = 0$ for any $n \in \mathbf{Z}$.*

PROOF If G is of order prime to p, then M is uniquely divisible by $|G|$, so $H^n(G, M) = 0$ for all $n \in \mathbf{Z}$ (15.8). Suppose p divides the order of G. Since G is p-nilpotent, there is a normal subgroup $S < G$ of index p. By induction assume our assertion true for S. The proof of our assertion is a repetition of the proof of 38.3.

39.2 Let G be a finite group. Let $P^i(G)$ be the statement:

$P^i(G)$: *If $H^n(G, M) = 0$ for some $n \in \mathbf{Z}$ and some G-module M of p-power order, then $H^{n+i}(G, M) = 0$*

for all $i \in \mathbf{Z}$.

We claim $P^i(G)$ implies $P^i(S)$ for every $S \leqq G$. For, suppose $H^n(S, M) = 0$ for some $S \leqq G$ and some $n \in \mathbf{Z}$. Then by Shapiro's lemma (34.2),

$$H^n(G, \mathbf{Z}G \otimes_{\mathbf{Z}S} M) \cong H^n(S, M) = 0. \tag{7}$$

Therefore, by $P^i(G)$ and Shapiro's lemma,

$$H^{n+i}(S, M) \cong H^{n+i}(G; \mathbf{Z}G \otimes_{\mathbf{Z}S} M) = 0.$$

Hence $P^i(S)$ holds.

39.3 *Let G be a finite group such that $P^i(G)$ holds; then G is p-nilpotent.*

PROOF The condition $P^i(G)$ certainly implies

$P^i(G)'$: *If M is a G-module of p-power order, then $H^n(G, M) \neq 0$ for some $n \in \mathbf{Z}$ implies $H^{n+i}(G, M) \neq 0$ for all i.*

Case 1. If the order of G is prime to p, then trivially G is p-nilpotent.

Case 2. If p divides the order of G, then

$$H^0(G, \mathbf{Z}/p\mathbf{Z}) \cong (\mathbf{Z}/p\mathbf{Z})^G/T_G(\mathbf{Z}/p\mathbf{Z}) = \mathbf{Z}/p\mathbf{Z} \neq 0.$$

Hence by $P^i(G)'$,

$$H^{-2}(G, \mathbf{Z}/p\mathbf{Z}) \cong \mathbf{Z}/p\mathbf{Z} \otimes G/[G, G] \neq 0,$$

by 21.2, and is isomorphic to the maximal abelian p-elementary factor group of G. Hence G contains a normal subgroup S of index p. By 39.2, $P^i(S)$ holds. Using induction suppose S is p-nilpotent with p-kernel K. Then K is a characteristic subgroup of S, and therefore is normal in G. Since the order of K is prime to p and its index in G is a p-power, G is p-nilpotent and K is its p-kernel.

39.4 *For a finite group G the following conditions are equivalent:*

(i) *G is p-nilpotent.*

(ii) *If M is a finite G-module of p-power order and $H^n(G, M) = 0$ for some $n \in \mathbf{Z}$, then $H^r(G, M) = 0$ for all $r \in \mathbf{Z}$.*

PROOF Condition (i) implies (ii) by 39.1 and dimension shifting; (ii) implies (i) by 39.3.

39.5 If G is finite nilpotent and M is a finite G-module, we have $M = \sum_p M_p$ where M_p is the p-primary component of M. Moreover, for any $S \leqq G$,

$$H^r(S, M) \cong \sum_p H^r(S, M_p).$$

Therefore $H^r(S, M) = 0$ if and only if $H^r(S, M_p) = 0$ for every p. Thus 39.4 implies that:

For a finite group G the following conditions are equivalent:

(i) *G is nilpotent.*
(ii) *If M is a finite G-module and $H^n(G, M) = 0$ for some $n \in \mathbf{Z}$, then $H^r(G, M) = 0$ for all $r \in \mathbf{Z}$.*

40. Stable Elements

In this section we follow Cartan–Eilenberg [15] to obtain the double coset formula (40.2 below) and the result contained in 40.4 needed in the next section.

Let $S, S' \leqq G$. For $g \in G$, the set of elements

$$SgS' = \{sgs' : s \in S, s' \in S'\}$$

is called a *double coset* with respect to S and S'. If two double cosets SgS' and ShS' have an element x in common, they are identical. For, let $x = sgs' = tht'$ with $s, t \in S$ and $s', t' \in S'$. If $x_1 \in SgS'$, then

$$x_1 = s_1 g s_1' = s_1 s^{-1} x s'^{-1} s_1' = s_1 s^{-1} t h t' s'^{-1} s_1' \in ShS'.$$

So $SgS' \subseteqq ShS'$. Similarly, $SgS' \supseteqq ShS'$.

40.1 *Let S and S' be subgroups of G. Then G can be written as the disjoint union $G = \dot{\bigcup}_i Sg_iS'$ of double cosets of S and S'. Moreover*

$$(G: S') = \sum_i (S: S \cap g_i S' g_i^{-1}). \tag{8}$$

PROOF $G = \dot{\bigcup}_i Sg_iS'$ because two double cosets are either equal or disjoint. To verify the identity (8), let $U_i = S \cap g_iS'g_i^{-1}$ and $S = \dot{\bigcup}_j s_{ij}U_i$. Then

$$Sg_i = \dot{\bigcup_j} s_{ij}(Sg_i \cap g_iS').$$

Hence

$$Sg_iS' = \dot{\bigcup_j} s_{ij}(Sg_iS' \cap g_iS') = \dot{\bigcup_j} s_{ij}g_iS'.$$

Thus

$$G = \dot{\bigcup_i} Sg_iS' = \dot{\bigcup_{i,j}} s_{ij}g_iS'. \tag{9}$$

This implies (8).

40.2 (*The double coset formula.*) *Let* $G = \dot{\bigcup}_i Sg_iS'$ *be a double coset representation of* G. *Then*

$$\mathrm{res}_{G,S}\,\mathrm{cor}_{S',G} = \sum_i \mathrm{cor}_{S\cap g_iS'g_i^{-1},S}\,\mathrm{res}_{g_iS'g_i^{-1},S\cap g_iS'g_i^{-1}}\,c_{g_i} \tag{10}$$

where c_{g_i}: $H^r(S', M) \to H^r(g_iS'g_i^{-1}, M)$ *is induced by the map* c_{g_i}: $\mathrm{Hom}_{S'}(X_r, M) \to \mathrm{Hom}_{g_iS'g_i^{-1}}(X_r, M)$ *where* X_* *is a complete resolution of* G, *and* $c_{g_i}\varphi = g_i\varphi g_i^{-1}$.

PROOF Let $\varphi \in \mathrm{Hom}_{S'}(X_r, M)$. Then by (9),

$$\begin{aligned}\mathrm{res}_{G,S}\,\mathrm{cor}_{S',G}\,\varphi &= \sum_{i,j} \varphi^{s_{i,j}g_i}\\ &= \sum_{i,j} c_{s_{ij}}c_{g_i}\varphi\\ &= \sum_i \mathrm{cor}_{S\cap g_iS'g_i^{-1},S}\,\mathrm{res}_{g_iS'g_i^{-1},S\cap g_iS'g_i^{-1}}\,c_{g_i}\varphi.\end{aligned}$$

An element $t \in H^r(S, M)$ is called *stable* if for each $g \in G$,

$$\mathrm{res}^r_{S,S\cap gSg^{-1}}\,t = \mathrm{res}^r_{gSg^{-1},S\cap gSg^{-1}}\,c_gt. \tag{11}$$

If $S \trianglelefteq G$ then, by (11), $t \in H^r(S, M)$ is stable if $t = t^g$ for every $g \in G$; i.e., t is stable if t is fixed under the action of G/S.

40.3 *If $t \in H^r(S, M)$ satisfies $t = \operatorname{res}_{G,S} t'$ for some $t' \in H^r(G, M)$, then t is stable.*

PROOF $c_g t' = gt'g^{-1} = t'$ since t' is a G-homomorphism. Hence $c_g t = c_g \operatorname{res}_{G,S} t' = \operatorname{res}_{G,gSg^{-1}} c_g t' = \operatorname{res}_{G,gSg^{-1}} t'$.
Therefore

$$\begin{aligned}\operatorname{res}_{gSg^{-1},S\cap gSg^{-1}} c_g t &= \operatorname{res}_{gSg^{-1},S\cap gSg^{-1}} \operatorname{res}_{G,gSg^{-1}} t' \\ &= \operatorname{res}_{G,S\cap gSg^{-1}} t' \\ &= \operatorname{res}_{S,S\cap gSg^{-1}} \operatorname{res}_{G,S} t' \\ &= \operatorname{res}_{S,S\cap gSg^{-1}} t.\end{aligned}$$

40.4 *If $t \in H^r(S, M)$ is stable, then*

$$\operatorname{res}_{G,S} \operatorname{cor}_{S,G} t = (G\colon S)t$$

where $(G\colon S)$ is the index of S in G.

PROOF By the double coset formula (10) in the case $S' = S$ we have

$$\begin{aligned}\operatorname{res}_{G,S} \operatorname{cor}_{S,G} t &= \sum_i \operatorname{cor}_{S\cap g_iSg_i^{-1},S} \operatorname{res}_{g_iSg^{-1},S\cap g_iSg_i^{-1}} c_{g_i} t \\ &= \sum_i \operatorname{cor}_{S\cap g_iSg_i^{-1},S} \operatorname{res}_{S,S\cap g_iSg_i^{-1}} t \qquad \text{by (11)} \\ &= \sum_i (S\colon S \cap g_iSg_i^{-1})t \\ &= (G\colon S)t \qquad \text{by 40.1.}\end{aligned}$$

41. Frobenius Groups

A *Frobenius group* is a finite group G which contains a nontrivial normal subgroup K such that if $k \in K$, $k \neq 1$, then the centralizer of k is contained in K. The subgroup K is called the *Frobenius kernel* of G. (A Frobenius group has a unique Frobenius kernel; see Scott [92, Theorem 12.6. 12].)

41.1 We list here some of the properties of Frobenius groups needed in this section. The proofs of these statements can be found in Gorenstein [39], Huppert [56], and Scott [92].

Let G be a Frobenius group with Frobenius kernel K. Then:

(i) *K is a Hall subgroup of G.*

(ii) *There exists a Hall subgroup S such that $G = KS$, $K \cap S = 1$; i.e., G is a semidirect product of K by S.* (S is called a *complement* of K in G.)

(iii) *For any complement S and any $g \in G$, $g \notin S$, we have $S \cap gSg^{-1} = 1$.*

(iv) *$|S|$ divides $|K| - 1$.*

(v) *Every element $s \in S$, $s \neq 1$, induces by conjugation an automorphism of K which fixes only the identity element of K.*

(vi) Let $S \leqq G$, $S \neq \{1\}$. Then G is Frobenius with complement S if and only if S is disjoint from its conjugates and is its own normalizer in G.

We also state the following result whose proof is in Zassenhaus [117, p. 132].

41.2 (*The Schur splitting theorem.*) *Let H be a normal Hall subgroup of G. Then G contains a subgroup S such that G is a semidirect product of H by S.*

41.3 *Let M be a G-module and S a subgroup of index k in G. If M is uniquely divisible by k, then*

$$\mathrm{res}^r_{G,S}\colon H^r(G, M) \to H^r(S, M)$$

is

(i) *a monomorphism for all $r \in \mathbf{Z}$,*

(ii) *an isomorphism for all $r \in \mathbf{Z}$ if G is Frobenius with complement S.*

Proof of (i) Let X_* be a complete resolution of G. Let $t \in \mathrm{Hom}_G(X_r, M)$ be a cocycle representing an element $\bar{t} \in H^r(G, M)$. If $(1/k)1_M$ is the inverse of $k1_M$, then

$$\begin{aligned}\text{cor}_{S,G}\cdot\text{Hom}_S\left(X_r, \frac{1}{k}1_M\right)\cdot\text{res}_{G,S}\, t &= T_{S,G}\left(\frac{1}{k}1_M t\right)\\ &= T_{S,G}\left(\frac{1}{k}1_M\right)t \qquad \text{since} \quad t \text{ is a}\\ &\qquad G\text{-homomorphism}\\ &= t.\end{aligned}$$

Therefore $\text{res}^r_{G,S}$ is a monomorphism.

PROOF OF (ii) For any $t \in H^r(S, M)$ the identity (11) in (40.2) is satisfied since $S \cap gSg^{-1} = 1$ for $g \in G$, $g \notin S$, and $gSg^{-1} = S$, $c_g = 1_{H^r(S,M)}$ for $g \in S$. Hence the elements of $H^r(S, M)$ are stable. By 40.4,

$$\text{res}^r_{G,S}\,\text{cor}^r_{S,G}\, t = kt = t \qquad \text{since} \quad S \text{ divides } k - 1.$$

Therefore $\text{res}^r_{G,S}$ is an epimorphism. This by (i) implies (ii).

41.4 *Let $\bar{G}$ be a finite p-group and M a $\bar{G}$-module of p-power order. If $H^n(\bar{G}, M) = 0$ for some $n \in \mathbf{Z}$, then M is cohomologically trivial* [*i.e., $H^r(\bar{S}, M) = 0$ for all $r \in \mathbf{Z}$ and all $\bar{S} \leqq \bar{G}$*].

PROOF This follows from Gaschütz's theorem 38.4, and 39.1.

41.5 *Let G be Frobenius with complement $\bar{G}$ of p-power order. Then for any finite G-module M of p-power order, $H^n(G, M) = 0$ for some $n \in \mathbf{Z}$ implies M is cohomologically trivially* [*i.e., $H^r(S, M) = 0$ for all $r \in \mathbf{Z}$ and all $S \leqq G$*].

PROOF *Case 1.* The order $|S|$ of S is prime to p. Then $H^r(S, M) = 0$ for all $r \in \mathbf{Z}$ since M is uniquely divisible by $|S|$.

Case 2. p divides $|S|$. Let K be the Frobenius kernel of G and $\bar{G}$ a complement of K. Since $K \cap S$ is normal in S and $S/K \cap S = KS/S$, it follows that $K \cap S$ is a normal Hall subgroup of S. By Schur's splitting theorem (41.2), $S \cap K$ has a complement $\bar{S}$ in S. Since $\bar{S}$ is of p-power order, $\bar{S} \leqq g\bar{G}g^{-1}$ for some $g \in G$. Our hypothesis, by 41.3 (ii), implies $H^n(g\bar{G}g^{-1}, M) = 0$. Moreover since $g\bar{G}g^{-1}$ is a p-group, by 41.4, $H^r(\bar{S},$

$M) = 0$ for all $r \in \mathbf{Z}$. Finally, since $\bar{S}$ is a subgroup of S of index prime to p, by 41.3, $H^r(S, M) = 0$ for all $r \in \mathbf{Z}$.

REMARK The result obtained in 41.5 does not hold if G is an arbitrary p-nilpotent group. For, let $G = \langle g \rangle$, $g^6 = 1$, $S = \langle g^2 \rangle$, $M = \langle m \rangle$, $3m = 0$, $gm = 2m$. Then G is 3-nilpotent, $|M| = 3$, $H^n(G, M) = 0$ for all $n \in \mathbf{Z}$ but $H^0(S, M) \neq 0$.

CHAPTER

7

Cup Products

42. Definition of Cup Product

42.1 Let

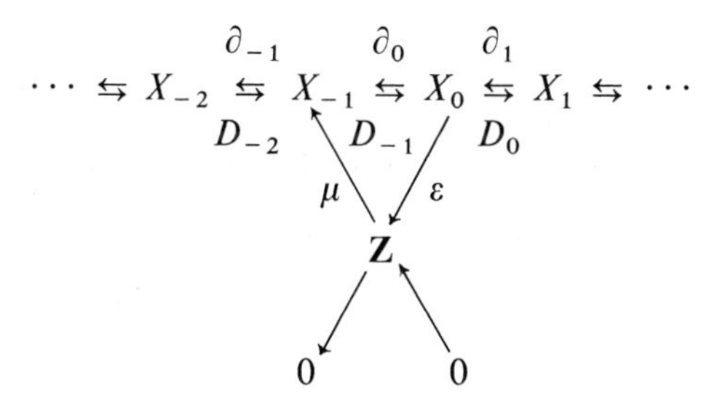

be a complete resolution of the finite group G. Recall that $\partial_{r+1}D_r + D_{r-1}\partial_r = 1_{X_r}$ for every r where D_r is a $\mathbf{Z}$-homomorphism. For each integer r we have a $\mathbf{Z}$-homomorphism $\pi_r \in \mathrm{Hom}_{\mathbf{Z}}(X_r, X_r)$ such that $T_G(\pi_r) = 1_{X_r}$. In the rest of this chapter we will omit writing the subscripts whenever no ambiguity ensues from such omissions. Define

$$\partial' = (\partial \otimes 1): X_p \otimes X_q \to X_{p-1} \otimes X_q$$

$$\partial'' = (1 \otimes \partial): X_p \otimes X_q \to X_p \otimes X_{q-1}$$
$$D' = T_G(D \otimes \pi): X_p \otimes X_q \to X_{p+1} \otimes X_q$$
$$D'' = T_G(\pi \otimes D): X_p \otimes X_q \to X_p \otimes X_{q+1}.$$

Thus D', D'' are G-homomorphisms (whereas D is a $\mathbf{Z}$-homomorphism); moreover

$$\partial' D' + D' \partial' = 1.$$

For,

$$\begin{aligned}
\partial' D' + D' \partial' &= T_G((\partial \otimes 1)(D \otimes \pi) + (D \otimes \pi)(\partial \otimes 1)) \\
&\qquad \text{since} \quad \partial \otimes 1 \text{ is a } G\text{-homomorphism} \\
&= T_G(\partial D \otimes \pi + D\partial \otimes \pi) \\
&= T_G(1 \otimes \pi) \\
&= \sum_{\sigma \in G} 1 \otimes \pi^\sigma \\
&= 1 \otimes T_G(\pi) \\
&= 1.
\end{aligned}$$

Similarly, $\partial'' D'' + D'' \partial'' = 1$. Therefore the G-homomorphisms D', D'' are homotopies.

42.2 *Let X_* be a complete resolution of G. There exist G-homomorphisms $\varphi_{p,q}: X_{p+q} \to X_p \otimes X_q$ for every $p, q \in \mathbf{Z}$ such that the identities $[p, q]$ and (1) hold:*

$[p, q]$: $$\varphi_{p,q}\partial = \partial' \varphi_{p+1,q} + (-1)^p \partial'' \varphi_{p,q+1}$$

(1) $$(\varepsilon \otimes \varepsilon)\varphi_{0,0} = \varepsilon$$

where $(\varepsilon \otimes \varepsilon)\varphi_{0,0}: X_0 \xrightarrow{\varphi_{0,0}} X_0 \otimes X_0 \xrightarrow{\varepsilon \otimes \varepsilon} \mathbf{Z} \otimes \mathbf{Z}\ (= \mathbf{Z})$.

PROOF Let $\partial'[p, q]$ be as follows:

$\partial'[p, q]$: $$\partial' \varphi_{p,q}\partial = (-1)^p \partial' \partial'' \varphi_{p,q+1}.$$

Then $[p, q] \to \partial'[p, q]$. Assume for one p and all q, $\varphi_{p,q}$ is defined and satisfies $\partial'[p, q]$. Then

$$\text{(2)}\qquad \begin{aligned} \varphi_{p,q}\partial &= (\partial' D' + D'\partial')\varphi_{p,q}\partial \\ &= \partial' D'\varphi_{p,q}\partial + (-1)^p(1 - \partial' D')\partial''\varphi_{p,q+1} \qquad \text{by } \partial'[p, q] \\ &= \partial'(D'\varphi_{p,q}\partial - (-1)^p D'\partial''\varphi_{p,q+1}) + (-1)^p\partial''\varphi_{p,q+1}. \end{aligned}$$

Define

$$\varphi_{p+1,q} = D'\varphi_{p,q}\partial + (-1)^{p+1}D'\partial''\varphi_{p,q+1}.$$

Thus (2) asserts that if for fixed p and all q, $\varphi_{p,q}$ is defined and satisfies $\partial'[p, q]$, then $\varphi_{p+1,q}[p,q]$ is defined and $[p, q]$ is satisfied. The computation of $[p, q]\partial$ yields

$$\partial'[p + 1, q]: \qquad \partial'\varphi_{p+1,q}\partial = (-1)^{p+1}\partial'\partial''\varphi_{p+1,q+1}.$$

Thus by induction we can define $\varphi_{s,q}$ for all $s \geqq p$ and all q such that $[s, q]$ is satisfied. Again assume for one p and all q, $\varphi_{p,q}$ is defined and satisfies $\partial'[p, q]$. Define

$$\text{(3)}\qquad \varphi_{p-1,q} = (-1)^p D''\partial'\varphi_{p,q-1}.$$

Then

$$\begin{aligned} \varphi_{p-1,q}\partial &= (-1)^{2p}D''\partial'\partial''\varphi_{p,q} \qquad \text{by } \partial'[p, q - 1] \\ &= (1 - \partial'' D'')\partial'\varphi_{p,q} \\ &= \partial'\varphi_{p,q} + (-1)^{p+1}\partial''\varphi_{p-1,q+1} \qquad \text{by (3)} \end{aligned}$$

which asserts $[p - 1, q]$ holds. Moreover the computation of $\partial'[p - 1, q]$ yields

$$\partial'[p - 1, q]: \qquad \partial'\varphi_{p-1,q}\partial = (-1)^{p-1}\partial'\partial''\varphi_{p-1,q+1}.$$

Thus by induction downward we can define $\varphi_{s,q}$ for all $s \leqq p$ and all q. It is enough, therefore, to construct $\varphi_{0,q}$ for all q such that $\partial'[0, q]$ and (1) hold. Choose an element $\xi \in X_0$ such that $\varepsilon(\xi) = 1$. For each q let $\Psi_{0,q}: X_q \to X_0 \otimes X_q$ be the **Z**-homomorphism defined by

$$\Psi_{0,q}(x) = \xi \otimes \pi_q(x).$$

Let

$$\varphi_{0,q} = T_G(\Psi_{0,q}).$$

Explicitly, the G-homomorphism $\varphi_{0,q}$ is given by

$$\varphi_{0,q}(x) = \sum_{\sigma \in G} \sigma \Psi_{0,q}(\sigma^{-1}x) = \sum_{\sigma \in G} \sigma\xi \otimes \pi_q^\sigma x.$$

$\varphi_{0,q}$ satisfies (1), for,

$$\begin{aligned}(\varepsilon \otimes \varepsilon)\varphi_{0,0}(x) &= \sum_{\sigma \in G} \varepsilon(\sigma\xi) \otimes \varepsilon\pi_q^\sigma x \\ &= 1 \otimes \varepsilon(\sum_{\sigma \in G} \pi_q^\sigma)x \\ &= 1 \otimes \varepsilon x \\ &= \varepsilon x \qquad \text{by the identification} \quad 1 \otimes z = z, \quad \text{for} \quad z \in \mathbf{Z}.\end{aligned}$$

To show $\partial'[0, q]$ is satisfied, let $x \in X_{q+1}$, then

$$\begin{aligned}\partial'\partial''\varphi_{0,q+1}(x) &= \partial''\partial'\varphi_{0,q+1}(x) \\ &= \partial''(\mu \otimes 1)(\varepsilon \otimes 1) \sum_{\sigma \in G} \sigma\xi \otimes \pi_{q+1}^\sigma x \\ &= \partial''(\mu \otimes 1)(1 \otimes \sum_{\sigma \in G} \pi_{q+1}^\sigma x) \\ &= (1 \otimes \partial)(\mu(1) \otimes x) \\ &= \mu(1) \otimes \partial x\end{aligned}$$

and

$$\begin{aligned}\partial'\varphi_{0,q}\partial x &= \partial' \sum_{\sigma \in G} \sigma\xi \otimes \pi_{q+1}^\sigma \partial x \\ &= (\mu \otimes 1)(1 \otimes \sum_{\sigma \in G} \pi_{q+1}^\sigma \partial x) \\ &= \mu(1) \otimes \partial x.\end{aligned}$$

42.3 Let $f \in \mathrm{Hom}_G(X_p, A)$, $g \in \mathrm{Hom}_G(X_q, B)$. Define the G-homomorphism

$$f \cup g : X_{p+q} \to A \otimes B$$

called the *cup product* by

$$f \cup g = (f \otimes g)\varphi_{p,q}.$$

43. Properties of Cup Products

43.1 *$f \cup g$ is linear in f and in g; moreover*

(4) $$\delta(f \cup g) = \delta f \cup g + (-1)^p f \cup \delta g$$

PROOF The bilinearity is obvious, and (4) is an identity since

$$\delta(f \cup g) = (f \otimes g)(\partial \otimes 1)\varphi_{p+1,q} + (-1)^p(f \otimes g)(1 \otimes \partial)\varphi_{p,q+1}$$

by $[p, q]$ in 42.2

$$= (f\partial \otimes g)\varphi_{p+1,q} + (-1)^p(f \otimes g\partial)\varphi_{p,q+1}$$

$$= \delta f \cup g + (-1)^p f \cup \delta g.$$

43.2 It follows from (4) that:

(i) cocycle $\cup$ cocycle = cocycle
(ii) coboundary $\cup$ cocycle = coboundary
(iii) cocycle $\cup$ coboundary = coboundary.

Moreover

(iv) *$a\varepsilon \cup b\varepsilon = (a \otimes b)\varepsilon$ for any $a \in A$, $b \in B$, where ε is the augmentation map.*
(v) *$\eta f \cup \theta g = (\eta \otimes \theta)(f \cup g)$ for any G-homomorphisms $\eta: A \to A'$, $\theta: B \to B'$.*

To show (iv), let $f_a: Z \to A$ be defined by $f_a(1) = a$, and write $a\varepsilon$ for $f_a \cdot \varepsilon$. Similarly, for $b \in B$ define f_b and write $b\varepsilon$ for $f_b \cdot \varepsilon$. Then

$$a\varepsilon \cup b\varepsilon = (f_a \otimes f_b)(\varepsilon \otimes \varepsilon)\varphi_{0,0} = (a \otimes b)\varepsilon \qquad \text{by (1).}$$

Condition (v) holds since

$$\eta f \cup \theta g = (\eta \otimes \theta)(f \otimes g)\varphi_{p,q} = (\eta \otimes \theta)(f \cup g).$$

43.3 Let A_ν be a G-module and $f_\nu \in \mathrm{Hom}_G(X_{p_\nu}, A_\nu)$, for $\nu = 1, 2, \ldots, m$.

Define

$$(5)\qquad f_1 \cup f_2 \cup \cdots \cup f_m = (\cdots((f_1 \cup f_2) \cup f_3)\cdots) \cup f_m).$$

Then $f_1 \cup f_2 \cup \cdots \cup f_m \in \mathrm{Hom}_G(X_{\Sigma p_\nu}, A_1 \otimes A_2 \otimes \cdots \otimes A_m)$. Moreover

(i) cup product is multilinear

(ii) $a_1\varepsilon \cup a_2\varepsilon \cup \cdots \cup a_m\varepsilon = (a_1 \otimes a_2 \otimes \cdots \otimes a_m)\varepsilon$

(iii) $\delta(f_1 \cup \cdots \cup f_m) = \sum_\nu (-1)^{p_1+\cdots+p_{\nu-1}} f_1 \cup \cdots \cup \delta f_\nu \cup \cdots \cup f_m$

(iv) $\eta_1 f_1 \cup \eta_2 f_2 \cup \cdots \cup \eta_m f_m$
$= (\eta_1 \otimes \eta_2 \otimes \cdots \otimes \eta_m)(f_1 \cup f_2 \cup \cdots \cup f_m)$.

It follows from (iii) that:

(v) If all f_ν are cocycles, then $f_1 \cup f_2 \cup \cdots \cup f_m$ is a cocycle.

(vi) If all f_ν are cocycles and $f_\iota = \delta g_\iota$, then

$$\delta(f_1 \cup \cdots \cup g_\iota \cup \cdots \cup f_m) = (-1)^{p_1+\cdots+p_{\iota-1}} f_1 \cup f_2 \cup \cdots \cup f_m.$$

Thus $f_1 \cup f_2 \cup \cdots \cup f_m$ is a coboundary.

44. Uniqueness

44.1 Let $\alpha_\nu \in H^{p_\nu}(G, A_\nu)$ be represented by the cocycle f_ν. Denote by $\alpha_1 \cup \alpha_2 \cup \cdots \cup \alpha_m$ the cocycle class in $H^{p_1+\cdots+p_m}(G, A_1 \otimes \cdots \otimes A_m)$ represented by $f_1 \cup f_2 \cup \cdots \cup f_m$. By 43.3, (v) and (vi), $\alpha_1 \cup \alpha_2 \cup \cdots \cup \alpha_m$ is well defined. Moreover

(i) Let $\kappa_0 \colon A_\iota^G \to H^0(G, A_\iota)$ be the map $\kappa_{0,G}^{A_\iota}$ defined in the proof of 37.3. For $a_\iota \in A_\iota^G$ we have

$$(6)\qquad \kappa_0(a_\iota) = a_\iota\varepsilon.$$

Thus

$$\begin{aligned}\kappa_0(a_1) \cup \cdots \cup \kappa_0(a_m) &= (a_1 \otimes \cdots \otimes a_m)\varepsilon \\ &= \kappa_0(a_1 \otimes \cdots \otimes a_m) \qquad \text{by (6).}\end{aligned}$$

This identity shows that $\cup$ is unique for dimension zero. We will show uniqueness for all dimensions in 44.2.

(ii) Let $\eta_\iota^*: H^{p_\iota}(G, A_\iota) \to H^{p_\iota}(G, A'_\iota)$ be the map induced by η_ι. Then $\eta_1^*\alpha_1 \cup \cdots \cup \eta_m^*\alpha_m = (\eta_1 \otimes \cdots \otimes \eta_m)(\alpha_1 \cup \cdots \cup \alpha_m)$.

(iii) Let $A_1, \ldots, A_m$ be G-modules, and for one of the indices ι, suppose $0 \to A_\iota \xrightarrow{i} A'_\iota \xrightarrow{j} A''_\iota \to 0$ is exact for some G-modules A'_ι, A''_ι. If

$$0 \to A_1 \otimes \cdots \otimes A_m \xrightarrow{1 \otimes \cdots \otimes i \otimes \cdots \otimes 1} A_1 \otimes \cdots \otimes A'_\iota \otimes \cdots \otimes A_m$$

$$\xrightarrow{1 \otimes \cdots \otimes j \otimes \cdots \otimes 1} A_1 \otimes \cdots \otimes A''_\iota \otimes \cdots \otimes A_m \to 0$$

is exact, then

$$\begin{array}{ccc} H^{p_1}(G, A_1) \times \cdots \times H^{p_\iota}(G, A_\iota) \times \cdots \times H^{p_m}(G, A_m) & \xrightarrow{\cup} & H^{\Sigma_{v=1}^m p_v}(G, A_1 \otimes \cdots \otimes A''_\iota \otimes \cdots \otimes A_m) \\ \downarrow 1 \times \cdots \times \delta^* \times \cdots \times 1 & & \downarrow \delta^* \\ H^{p_1}(G, A_1) \times \cdots \times H^{p_\iota+1}(G, A''_\iota) \times \cdots \times H^{p_m}(G, A_m) & \xrightarrow{\cup} & H^{\Sigma_{v=1}^m p_v+1}(G, A_1 \otimes \cdots \otimes A_\iota \otimes \cdots \otimes A_m) \end{array}$$

is commutative up to the sign $(-1)^{p_1+\cdots+p_{\iota-1}}$.

To show (iii) let $f_{p_v} \in \mathrm{Hom}_G(X_{p_v}, A_v)$ represent $\alpha_{p_v} \in H^{p_v}(G, A_v)$, $v \neq \iota$, and $f''_{p_\iota} \in \mathrm{Hom}_G(X_{p_\iota}, A''_\iota)$ represent $\alpha''_{p_\iota} \in H^{p_\iota}(G, A''_\iota)$. To compute $\delta^*\alpha''_{p_\iota}$ first write $f''_{p_\iota} = jf'_{p_\iota}$, then $\delta f'_{p_\iota} = f'_{p_\iota}\partial_{p_\iota+1} = if_{p_\iota+1}$. By 8.3, $\delta^*\alpha''_{p_\iota}$ is represented by the cocycle $f_{p_\iota+1}$. Thus in the diagram the class of $(f_1, \ldots, f''_\iota, \ldots, f_m)$ is carried counterclockwise to the class of $f_1 \cup \cdots \cup f_\iota \cup \cdots \cup f_m$. However, the same class is carried clockwise first to the class of

$$f_1 \cup \cdots \cup f''_\iota \cup \cdots \cup f_m$$
$$= (1 \otimes \cdots \otimes j \otimes \cdots \otimes 1)(f_1 \cup \cdots \cup f'_\iota \cup \cdots \cup f_m)$$

in $H^{\Sigma_{v=1}^{\iota-1} p_v}(G, A_1 \cdots A''_\iota \cdots A_m)$ and then by δ^* to the class whose image under $1 \otimes \cdots \otimes i \otimes \cdots \otimes 1$ is

$$(\delta(f_1 \cup \cdots \cup f'_\iota \cup \cdots \cup f_m)$$
$$= (-1)^{\Sigma_{v=1}^{\iota-1} p_v} f_1 \cup \cdots \cup \delta f'_\iota \cup \cdots \cup f_m$$
$$= (-1)^{\Sigma_{v=1}^{\iota-1} p_v} (1 \otimes \cdots \otimes \iota \otimes \cdots \otimes 1)(f_1 \cup \cdots \cup f_m).$$

Thus clockwise we get the class of $(-1)^{\Sigma_{v=1}^{\iota-1} p_v} f_1 \cup \cdots \cup f_m$.

44.2 *If $\cup$ satisfies the properties (i), (ii), and (iii) of* 44.1, *then it is unique.*

PROOF Property (i) gives the uniqueness of the cup product map

$$H^0(G, A_1) \times \cdots \times H^0(G, A_m) \xrightarrow{\cup} H^0(G, A_1 \otimes \cdots \otimes A_m).$$

I. (Induction upward.) Suppose for any G-modules $B_1, \ldots, B_m$ and integers $p_1, \ldots, p_m$ the cup product map

$$H^{p_1}(G, B_1) \times \cdots \times H^{p_m}(G, B_m) \xrightarrow{\cup} H^{p_1+\cdots+p_m}(G, B_1 \otimes \cdots \otimes B_m)$$

is unique. Consider the split exact sequence

$$0 \to Z \to \mathbf{Z}G \to J \to 0$$

of 17.1, (1b). Let A_ι be a G-module and $A'_\iota = \mathbf{Z}G \otimes A_\iota$, $A''_\iota = J \otimes A_\iota$. Then the G-regularity of $\mathbf{Z}G$ implies A'_ι is G-regular (or G-induced). Since

$$0 \to A_\iota \to A'_\iota \to A''_\iota \to 0$$

is exact (17.1,(2)), for A_ι, A'_ι, A''_ι and any G-modules $A_1, \ldots, A_{\iota-1}$, $A_{\iota+1}, \ldots, A_m$, the condition 44.1(iii) on the diagram is satisfied, and moreover the vertical maps there are isomorphisms. Thus the lower cup product map in the diagram of 44.1,(iii) is unique since by the induction hypothesis the upper cup product map is unique.

II. (Induction downward.) Use the exact sequence

$$0 \to I \to \mathbf{Z}G \to \mathbf{Z} \to 0$$

of 17.1 instead of $0 \to \mathbf{Z} \to \mathbf{Z}G \to J \to 0$. The procedure is the same as in I.

44.3 We remark that 44.2 shows the independence of the cup product from the choice of the complete resolution of G. For, if X_*, X'_* are complete resolutions of G, then by Section 42, they define two cup products

$$\cup, \cup': H^{p_1}(G, A_1) \times \cdots \times H^{p_m}(G, A_m) \to H^{\sum_{\nu=1}^m p_\nu}(G, A_1 \otimes \cdots \otimes A_m).$$

Since both $\cup$ and $\cup'$ satisfy (i), (ii), (iii) of 44.1 it follows from 44.2 that $\cup = \cup'$.

45. Further Properties of Cup Products

45.1 *Associativity.* Define the cup product

$$H^{p_1}(G, A_1) \times H^{p_2}(G, A_2) \times H^{p_3}(G, A_3) \to H^{p_1+p_2+p_3}(G, A_1 \otimes A_2 \otimes A_3)$$

by $(\alpha_1 \cup \alpha_2) \cup \alpha_3$ as we have done so far, and alternatively by $\alpha_1 \cup (\alpha_2 \cup \alpha_3)$. By the uniqueness of cup products

$$(\alpha_1 \cup \alpha_2) \cup \alpha_3 = \alpha_1 \cup (\alpha_2 \cup \alpha_3).$$

45.2 *Anticommutativity formula.* There exists an isomorphism

$$\tau: B \otimes A \xrightarrow{\sim} A \otimes B$$

defined by $\tau(b \otimes a) = a \otimes b$. For $\alpha \in H^p(G, A)$ and $\beta \in H^q(G, B)$ how is $\alpha \cup \beta$ related to $\beta \cup \alpha$? It will turn out that

$$\alpha \cup \beta = (-1)^{pq}\tau^*(\beta \cup \alpha).$$

Define a new map

$$\cup': \mathrm{Hom}_G(X_p, A) \times \mathrm{Hom}_G(X_q, B) \to \mathrm{Hom}(X_{p+q}, A \otimes B)$$

by

$$f \cup' g = (-1)^{pq}\tau(g \cup f).$$

Then $\delta(f \cup' g) = (-1)^{pq}\tau\delta(g \cup f)$ since τ induces a chain morphism

$$\tau: \mathrm{Hom}_G(X_{p+q}, B \otimes A) \to \mathrm{Hom}_G(X_{p+q}, A \otimes B),$$

and so

$$\begin{aligned}\delta(f \cup' g) &= (-1)^{(p+1)q}\tau(g \cup \delta f)\\ &\quad + (-1)^p(-1)^{p(q+1)}\tau(\delta g \cup f)\\ &= \delta f \cup' g + (-1)^p f \cup' \delta g.\end{aligned}$$

Thus the cochain $\cup'$ has the same differentiation formula as the cup $\cup$. Next, let $\eta: A \to A'$, $\theta: B \to B'$ be G-homomorphisms. Then

$$\eta f \cup' \theta g = (-1)^{pq}\tau(\theta g \cup \eta f) = (-1)^{pq}\tau(\theta \otimes \eta)(g \cup f)$$
$$= (-1)^{pq}(\eta \otimes \theta)\tau(g \cup f) = (\eta \otimes \theta)(f \cup' g).$$

Finally $a\varepsilon \cup' b\varepsilon = \tau(b\varepsilon \cup a\varepsilon) = \tau(b \otimes a)\varepsilon = (a \otimes b)\varepsilon$. Thus by the uniqueness of cup product $\alpha \cup' \beta = \alpha \cup \beta$. Therefore we have

$$\alpha \cup \beta = (-1)^{pq}\tau^*(\beta \cup \alpha),$$

which is the anticommutativity formula for cup products.

46. Pairings

46.1 Let A, B, and C be G-modules. Suppose a bilinear map $A \times B \xrightarrow{\cdot} C$ of **Z**-modules is given where we write $\cdot(a, b) = a \cdot b$. We say the map is a *G-pairing* of A and B to C if

$$\sigma(a \cdot b) = \sigma a \cdot \sigma b \qquad \text{for every} \quad \sigma \in G.$$

Thus if A, B, and C are viewed as **Z**-modules, there exists a **Z**-homomorphism $\Omega\colon A \otimes B \to C$ such that

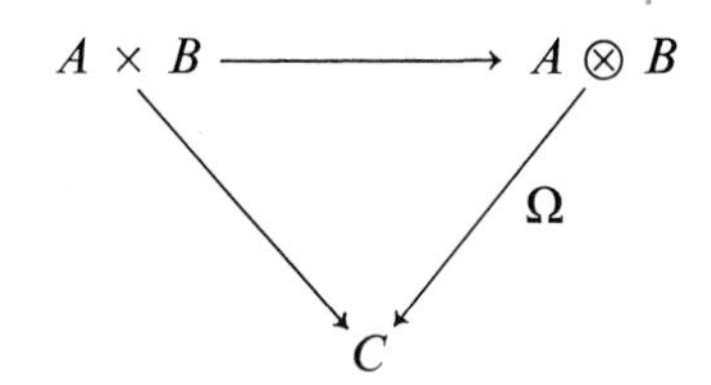

is commutative. Moreover $A \otimes B$ can be made into a G-module by $\sigma(a \otimes b) = \sigma a \otimes \sigma b$ and in this context Ω is a G-homomorphism, i.e.,

$$\sigma(\Omega(a \otimes b)) = \sigma(a \cdot b) = \sigma a \cdot \sigma b = \Omega(\sigma a \otimes \sigma b) = \Omega(\sigma(a \otimes b)).$$

Therefore Ω induces the homomorphism

$$H^p(G, A) \times H^q(G, B) \xrightarrow{\cup} H^{p+q}(G, A \otimes B) \xrightarrow{\Omega^*} H^{p+q}(G, C).$$

For $\alpha \in H^p(G, A)$, $\beta \in H^q(G, B)$, denote

$$\alpha \cdot \beta = \Omega^*(\alpha \cup \beta) \in H^{p+q}(G, C).$$

Since $\alpha \cup \beta$ is bilinear, so is $\alpha \cdot \beta$.

46.2 *Associativity.* Let A_1, A_2, A_3, A_{12}, A_{23}, A_{123} be G-modules. Suppose

$$A_1 \times A_2 \to A_{12} \qquad A_2 \times A_3 \to A_{23}$$
$$A_{12} \times A_3 \to A_{123} \qquad A_1 \times A_{23} \to A_{123}$$

are G-pairings such that $(a_1 \cdot a_2) \cdot a_3 = a_1 \cdot (a_2 \cdot a_3)$ where $a_1 \in A_1$, $a_2 \in A_2$, $a_3 \in A_3$. Then the diagram

$$\begin{array}{ccc} A_1 \otimes A_2 \otimes A_3 & \xrightarrow{\Omega_{12} \otimes 1} & A_{12} \otimes A_3 \\ \Big\downarrow {\scriptstyle 1 \otimes \Omega_{23}} & & \Big\downarrow {\scriptstyle \Omega_{12,3}} \\ A_1 \otimes A_{23} & \xrightarrow{\Omega_{1,23}} & A_{123} \end{array}$$

is commutative. In fact,

$$\Omega_{12,3}(\Omega_{12} \otimes 1)a_1 \otimes a_2 \otimes a_3 = \Omega_{12,3}((a_1 \cdot a_2) \otimes a_3) = (a_1 \cdot a_2) \cdot a_3$$

and

$$\Omega_{1,23}(1 \otimes \Omega_{23})a_1 \otimes a_2 \otimes a_3 = \Omega_{1,23}(a_1 \otimes (a_2 \cdot a_3)) = a_1 \cdot (a_2 \cdot a_3).$$

Let $\alpha_i \in H^{p_i}(G, A_i)$, $i = 1, 2, 3$. Then

$$\begin{aligned} (\alpha_1 \cdot \alpha_2) \cdot \alpha_3 &= \Omega^*_{12,3}(\Omega^*_{12}(\alpha_1 \cup \alpha_2) \cup \alpha_3) \\ &= (\Omega_{12,3}(\Omega_{12} \otimes 1))^*((\alpha_1 \cup \alpha_2) \cup \alpha_3) \\ &= (\Omega_{1,23}(1 \otimes \Omega_{2,3}))^*(\alpha_1 \cup (\alpha_2 \cup \alpha_3)) \\ &= \Omega^*_{1,23}(\alpha_1 \cup \Omega^*_{23}(\alpha_2 \cup \alpha_3)) \\ &= \Omega^*_{1,23}(\alpha_1 \cup (\alpha_2 \cdot \alpha_3)) \\ &= \alpha_1 \cdot (\alpha_2 \cdot \alpha_3). \end{aligned}$$

46.3 *Anticommutativity. Let $A \times B \xrightarrow{\cdot} C$ be a pairing. Then $\alpha \cdot \beta = (-1)^{pq} \beta \cdot \alpha$, where $\alpha \in H^p(G, A)$ and $\beta \in H^q(G, B)$.*

PROOF Define the pairing $B \times A \xrightarrow{\circ} C$ by $b \circ a = a \cdot b$. Then $a \cdot b = \Omega(a \otimes b)$ and $b \circ a = \Omega\tau(b \otimes a)$ where $\tau: B \otimes A \xrightarrow{\sim} A \otimes B$ is the isomorphism given in 45.2. Thus

$$\begin{aligned}\beta\cdot\alpha &= \Omega^*\tau^*(\beta\cup\alpha)\\ &= \Omega^*(-1)^{pq}\alpha\cup\beta \qquad \text{by 45.2}\\ &= (-1)^{pq}\alpha\cdot\beta.\end{aligned}$$

47. Duality

47.1 *Let W, X, Y be abelian groups. Then there exists a one-to-one onto map*

$$\alpha\colon \operatorname{Hom}(W\otimes X, Y) \rightrightarrows \operatorname{Hom}(W, \operatorname{Hom}(X, Y)).$$

α is in fact an isomorphism of abelian groups and is functoral in W, X, and Y.

PROOF For $\bar{f}\in \operatorname{Hom}(W\otimes X, Y)$ let $f\colon W\times X\to Y$ be the associated bilinear map. Define α by $(\alpha\bar{f})_w(x) = f(w, x)$. The bilinearity of f in w (resp. x) implies $\alpha\bar{f}$ (resp. $(\alpha\bar{f})_w$) is a homomorphism of abelian groups. For each $g\in \operatorname{Hom}(W, \operatorname{Hom}(X, Y))$ define the map

$$(\beta g)\colon W\times X\to Y$$

by $(\beta g)(w, x) = g_w(x)$; (βg) is bilinear since g and $g_w\colon X\to Y$ are homomorphisms of abelian groups. Let $(\tilde{\beta}g)\colon W\otimes X\to Y$ be the homomorphism induced by the bilinear map (βg). Define

$$\tilde{\beta}\colon \operatorname{Hom}(W, \operatorname{Hom}(X, Y))\to \operatorname{Hom}(W\otimes X, Y)$$

by $\tilde{\beta}(g) = (\tilde{\beta}g)$; i.e., $(\tilde{\beta}(g))(w\otimes x) = g_w(x)$. Then

$$(\tilde{\beta}(\alpha\bar{f}))(w\otimes x) = (\alpha\bar{f})_w(x) = \bar{f}(w\otimes x)$$

and

$$(\alpha(\tilde{\beta}g))_w(x) = (\tilde{\beta}g)(w\otimes x) = g_w(x).$$

47.2 Let B and C be G-modules. The map $G\times \operatorname{Hom}_{\mathbf{Z}}(B, C)\overset{\varphi}{\to} \operatorname{Hom}_{\mathbf{Z}}(B, C)$ defined by $\varphi(\sigma, f) = f^\sigma$ imposes a G-module structure on

$\mathrm{Hom}_{\mathbf{Z}}(B, C)$ (12.3). Then the evaluation map

$$\text{eval.}: \mathrm{Hom}_{\mathbf{Z}}(B, C) \times B \to C$$

defined by eval. $(f, b) = f(b)$ is a G-pairing. For eval. is a bilinear map of **Z**-modules; moreover

$$\sigma f(b) = \sigma f(\sigma^{-1}\sigma b) = f^{\sigma}(\sigma b).$$

By 46.1, eval. induces a unique bilinear map

$$H^p(G, \mathrm{Hom}(B, C)) \times H^q(G, B) \xrightarrow{\cdot} H^{p+q}(G, C).$$

Hence by the universal mapping property for tensor products eval. induces a unique group homomorphism

$$\gamma: H^p(G, \mathrm{Hom}(B, C)) \otimes H^q(G, B) \to H^{p+q}(G, C)$$

where, as in 46.1, we denote $\gamma(\alpha \otimes \beta)$ by $\alpha \cdot \beta$. Finally, by the isomorphism α of 47.1, eval. induces the unique homomorphism

$$\alpha\gamma = h_{p,q}: H^p(G, \mathrm{Hom}(B, C)) \to \mathrm{Hom}(H^q(G, B), H^{p+q}(G, C)).$$

Explicitly,

$$(h_{p,q}\varphi)(\beta) = \gamma(\varphi \otimes \beta) = \varphi \cdot \beta$$

for $\varphi \in H^p(G, \mathrm{Hom}(B, C))$, $\beta \in H^q(G, B)$.

47.3 *Let* $0 \to B' \xrightarrow{i} B \xrightarrow{j} B'' \to 0$ *be an exact sequence of* G*-homomorphisms. Let* C *be a* G*-module and suppose*

$$0 \to \mathrm{Hom}(B'', C) \xrightarrow{j'} \mathrm{Hom}(B, C) \xrightarrow{i'} \mathrm{Hom}(B', C) \to 0$$

is exact, where $j' = \mathrm{Hom}(j, 1_C)$ *and* $i' = \mathrm{Hom}(i, 1_C)$. *Let* $\varphi' \in H^p(G, \mathrm{Hom}(B', C))$ *and* $\beta'' \in H^{q-1}(G, B'')$. *Then* $\delta^*\varphi' \in H^{p+1}(G, \mathrm{Hom}(B'', C))$ *and* $\delta^*\beta'' \in H^q(G, B')$; *moreover*

$$\delta^*\varphi' \cdot \beta'' + (-1)^p \varphi' \cdot \delta^*\beta'' = 0. \tag{7}$$

Proof Let X_* be a complete resolution of G. Let the cocycle $F': X_p \to \mathrm{Hom}(B', C)$ represent φ', and $g'': X_q \to B''$ represent β''. Let $g: X_q \to B$ be such that $jg = g''$, and $g': X_{q+1} \to B'$ be such that $ig' = \delta g$. Then the cocycle g' represents $\delta^*\beta''$. Similarly, let $F: X_p \to \mathrm{Hom}(B, C)$ be such that $i'F = F'$, and $F'': X_{p+1} \to \mathrm{Hom}(B'', C)$ be such that $j'F'' = \delta F$. Then the cocycle F'' represents $\delta^*\varphi'$. Thus $(\delta^*\varphi')\cdot\beta'' + (-1)^p\varphi'\delta^*\beta'$ is represented by

$$\begin{aligned} F''\cdot g'' + (-1)^p F'\cdot g' &= F''\cdot jg + (-1)^p i'F\cdot g' \\ &= j'F''\cdot g + (-1)^p F\cdot ig' \\ &= \delta F\cdot g + (-1)^p F\cdot \delta g \\ &= \delta(F\cdot g) \end{aligned}$$

which implies (7). In the computation above we have used the identities

$$F''\cdot jg = j'F''\cdot g \qquad \text{and} \qquad i'F\cdot g' = G\cdot ig',$$

the verification of which is left to the reader.

47.4 Since, by 47.2,

$$(h_{p,q}\varphi')\delta^*\beta'' = \varphi'\cdot\delta^*\beta'' \qquad \text{and} \qquad (h_{p+1,q-1}\delta^*\varphi')\beta'' = \delta^*\varphi'\cdot\beta'',$$

the identity (7) of 47.3 states that the diagram

$$\begin{array}{ccc} H^p(G, \mathrm{Hom}(B', C)) & \xrightarrow{h_{p,q}} & \mathrm{Hom}(H^q(G, B'), H^{p+q}(G, C)) \\ \Big\downarrow \delta^* & & \Big\downarrow \mathrm{Hom}(\delta^*, 1) \\ H^{p+1}(G, \mathrm{Hom}(B'', C)) & \xrightarrow{h_{p+1,q-1}} & \mathrm{Hom}(H^{q-1}(G, B''), H^{p+q}(G, C)) \end{array}$$

commutes up to the sign $(-1)^{p+1}$.

47.5 *If $h_{p,q}$ is an isomorphism, so is $h_{p+1,q-1}$.*

PROOF Consider the split exact sequence $0 \to I \to \mathbf{Z}G \to \mathbf{Z} \to 0$. Since $\mathbf{Z}G$ is G-induced, the sequence $0 \to I \otimes B \to \mathbf{Z}G \otimes B \to B \to 0$ is split exact and $\mathbf{Z}G \otimes B$ is G-induced. Furthermore

$$0 \to \mathrm{Hom}_{\mathbf{Z}}(B, C) \to \mathrm{Hom}_{\mathbf{Z}}(\mathbf{Z}G \otimes B, C) \to \mathrm{Hom}_{\mathbf{Z}}(I \otimes B, C) \to 0$$

is exact and $\mathrm{Hom}_{\mathbf{Z}}(\mathbf{Z}G \otimes B, C)$ is G-induced (13.3). By 47.4, we have the commutative diagram

$$\begin{array}{ccc} H^p(G, \mathrm{Hom}(I \otimes B, C)) & \xrightarrow{h_{p,q}} & \mathrm{Hom}(H^q(G, I \otimes B), H^{p+q}(G, C)) \\ \downarrow \wr & & \downarrow \wr \\ H^{p+1}(G, \mathrm{Hom}(B, C)) & \xrightarrow{h_{p+1,q-1}} & \mathrm{Hom}(H^{q-1}(G, B), H^{p+q}(G, C)) \end{array}$$

where the vertical maps are isomorphisms since $\mathrm{Hom}(\mathbf{Z}G \otimes B, C)$ and $\mathbf{Z}G \otimes B$ are G-induced. Since $h_{p,q}$ is an isomorphism, so is $h_{p+1,q-1}$.

47.6 *If $h_{p,q}$ is an isomorphism, so is $h_{r,s}$ whenever $r + s = p + q$.*

Proof For $r \geqq p$ the assertion follows from 47.5 by induction. For $r < p$ repeat the proof of 47.5 starting with the split exact sequence $0 \to Z \to \mathbf{Z}G \to J \to 0$.

47.7 *Let B be an abelian group. Let*

$$\hat{B} = \mathrm{Hom}_{\mathbf{Z}}(B, \mathbf{Q}/\mathbf{Z}) = \textit{Group of characters of } B,$$

where $\mathbf{Q}$ is the rationals. If $\chi \in \mathrm{Hom}_{\mathbf{Z}}(B_0, \mathbf{Q}/\mathbf{Z})$ for any $B_0 \leqq B$, then χ can be extended to B.

Proof The extension is possible since $\mathbf{Q}/\mathbf{Z}$ is a divisible (injective) group. We show the existence of an extension of χ directly as follows: If $b \in B - B_0$, let B_1 be the group generated by the elements $zb + b_0$ where $z \in \mathbf{Z}$ and $b_0 \in B$. Let e be the smallest nonnegative integer such that $eb = b_1 \in B_0$. Extend χ to B_1 by letting $\chi(b) = (1/e)\chi(b_1)$ if $e \neq 0$, and $\chi(b)$ equal to any element of $\mathbf{Q}/\mathbf{Z}$ if $e = 0$. By Zorn's lemma χ can be extended to B.

47.8 $h_{0,-1}\colon H^0(G, \mathrm{Hom}(B, \mathbf{Q}/\mathbf{Z})) \to \mathrm{Hom}(H^{-1}(G, B), H^{-1}(G, \mathbf{Q}/\mathbf{Z}))$ *is an isomorphism.*

Proof To show $h_{0,-1}$ is onto, let $\varphi \in \mathrm{Hom}(H^{-1}(G, B), H^{-1}(G, \mathbf{Q}/\mathbf{Z}))$. Since $H^{-1}(G, \mathbf{Q}/\mathbf{Z}) \cong (1/n)\mathbf{Z}/\mathbf{Z}$ by 24,(viii), and $(1/n)\mathbf{Z}/\mathbf{Z}$ may be viewed as a subgroup of $\mathbf{Q}/\mathbf{Z}$, we can suppose $\varphi \in \text{Hom}(H^{-1}(G, B), \mathbf{Q}/\mathbf{Z})$. It is enough to show φ is induced by cupping followed by pairing. Since $H^{-1}(G, B) \cong B_T/IB$ where $B_T = \{\beta \in B \text{ s.t. } T(\beta) = 0\}$, φ induces $f_0\colon B_T \to \mathbf{Q}/\mathbf{Z}$ such that $f_0(IB) = 0$. Extend f_0 to $f\colon B \to \mathbf{Q}/\mathbf{Z}$ by 44.7. Then f is a G-homomorphism, for,

$$f(\sigma b) - \sigma f(b) = f(\sigma b) - f(b) = f(\sigma b - b) = 0 \qquad \text{since} \quad \sigma b - b \in IB.$$

Thus $f \in (\text{Hom}(B, \mathbf{Q}/\mathbf{Z}))^G = \hat{B}^G$. Hence $\kappa_0(f) \in H^0(G, \hat{B})$ where $\kappa_0\colon \hat{B}^G \to \hat{B}^G/T(\hat{B}) = H^0(G, \hat{B})$ is the canonical epimorphism. Let $\beta \in H^{-1}(G, B)$. Then for some $b \in B_T$, $\beta = \kappa_{-1}(b)$ where $\kappa_{-1}\colon B_T \to B_T/IB = H^{-1}(G, B)$ is the canonical epimorphism. We claim

$$\kappa_0(f)\cdot\kappa_{-1}(b) = \kappa_{-1}(fb). \tag{8}$$

To show this we observe that the left side is the image of $(\kappa_0(f), \kappa_{-1}(b))$ by the composite map

$$H^p(G, \text{Hom}(B, C)) \times H^q(G, B) \xrightarrow{\cup} H^{p+q}(G, \text{Hom}(B, C) \otimes B)$$
$$\xrightarrow{\overline{\text{eval}}^*} H^{p+q}(G, C)$$

where $\overline{\text{eval}}\colon \text{Hom}(B, C) \otimes B \to C$ is the homomorphism induced by eval. Let the cocycle $g\colon X_{-1} \to B$ represent $\kappa_{-1}(b)$, then $f\cdot g = fg$. Thus

$$\kappa_0(f)\cdot\kappa_{-1}(b) = \text{Class of } fg = \kappa_{-1}(fb) = \kappa_{-1}(f_0 b) = \varphi(\beta).$$

Thus $h_{0,-1}$ is onto. To show $h_{0,-1}$ is a monomorphism, suppose $f \in \text{Hom}_G(B, \mathbf{Q}/\mathbf{Z})$ and $\kappa_0(f)\cdot\beta = 0$ for every $\beta \in H^{-1}(G, B)$. So for $b \in B_T$, $0 = \kappa_0(f)\cdot\kappa_{-1}(b) = fb \in \mathbf{Q}/\mathbf{Z}$. Thus f vanishes on B_T. If $T(b_1) = T(b_2)$ for some $b_1, b_2 \in B$, then $f(b_1 - b_2) = 0$, and so $f(b_1) = f(b_2)$. Thus f is a class function and induces $g_0\colon T(B) \to \mathbf{Q}/\mathbf{Z}$, such that $f(b) = g_0 T(b)$. Extend g_0 to $g\colon B \to \mathbf{Q}/\mathbf{Z}$; then $T(g) = f$. Indeed

$$\begin{aligned}(T(g))(b) &= \sum g\sigma^{-1}b \qquad \text{since} \quad g\sigma^{-1}b \in \mathbf{Q}/\mathbf{Z} \\ &= gT(b) = g_0 T(b) = fb.\end{aligned}$$

So $\kappa_0(f) = 0$ since

$$H^0(G, \text{Hom}(B, \mathbf{Q}/\mathbf{Z}) \cong \text{Hom}_G(B, \mathbf{Q}/\mathbf{Z})/T(\text{Hom}(B, \mathbf{Q}/\mathbf{Z})).$$

47.9 As before $n = |G|$ and $H^{-1}(G, \mathbf{Q}/\mathbf{Z}) = (1/n)\mathbf{Z}/\mathbf{Z} \subset \mathbf{Q}/\mathbf{Z}$. Since the order of every element of $H^{q-1}(G, B)$ divides n, we have

$$\mathrm{Hom}(H^{q-1}(G, B), H^{-1}(G, \mathbf{Q}/\mathbf{Z}) \simeq (\mathrm{Hom}(H^{q-1}(G, B), \mathbf{Q}/\mathbf{Z})$$

where the isomorphism is induced by the identification $H^{-1}(G, \mathbf{Q}/\mathbf{Z}) = (1/n)\mathbf{Z}/\mathbf{Z}$ followed by the inclusion $(1/n)\mathbf{Z}/\mathbf{Z} \subset \mathbf{Q}/\mathbf{Z}$. Since

$$\mathrm{Hom}(H^{q-1}(G, B), \mathbf{Q}/\mathbf{Z}) = \widehat{H^{q-1}(G, B),}$$

the dual group of $H^{q-1}(G, B)$, by 47.6 and 47.8 we have the isomorphism

$$h_{-q,q-1}\colon H^{-q}(G, \hat{B}) \simeq \widehat{H^{q-1}(G, B).} \tag{9}$$

We have shown that $H^{-q}(G, \hat{B})$ is the group of characters of $H^{q-1}(G, B)$.

47.10 In particular, if we set $B = \mathbf{Z}$ in (9), then $\hat{B} = \mathrm{Hom}(Z, \mathbf{Q}/\mathbf{Z}) = \mathbf{Q}/\mathbf{Z}$ and

$$h_{-q,q-1}\colon H^{-q}(G, \mathbf{Q}/\mathbf{Z}) \simeq \widehat{H^{q-1}(G, \mathbf{Z}).} \tag{10}$$

Since $0 \to \mathbf{Z} \to \mathbf{Q} \to \mathbf{Q}/\mathbf{Z} \to 0$ with the obvious maps is exact, and $\mathbf{Q}$ is uniquely divisible by $n = |G|$ [thus $H^{-q}(G, \mathbf{Q}) = 0$ for every $q \in \mathbf{Z}$ (15.8)], we have

$$H^{-q}(G, \mathbf{Q}/\mathbf{Z}) \overset{\delta^*}{\simeq} H^{-q+1}(G, \mathbf{Z}). \tag{11}$$

Combining (10) and (11) we have the isomorphisms

$$H^{-q}(G, \mathbf{Z}) \overset{\alpha}{\cong} \widehat{H^{q}(G, \mathbf{Z})} \tag{12}$$

and

$$H^{-q}(G, \mathbf{Q}/\mathbf{Z}) \cong \widehat{H^{q+2}(G, \mathbf{Q}/\mathbf{Z}).}$$

48. Periodicity

48.1 *Let*

$$\tilde{\beta}\colon \mathrm{Hom}(W, \mathrm{Hom}(X, Y)) \simeq \mathrm{Hom}(W \otimes X, Y)$$

be the one-to-one onto map of 47.1 where W, X, and Y are abelian groups. Suppose

(i) *X is cyclic of order n,*
(ii) *the order of any element of W or Y divides n,*
(iii) *there exists an isomorphism* $h \in \mathrm{Hom}(W, \mathrm{Hom}(X, Y))$.

Then $\tilde{\beta}h\colon W \otimes Y \to X$ *is an isomorphism.*

PROOF Without loss of generality we may assume $X = \mathbf{Z}/n\mathbf{Z}$. By (ii), W, X, and Y may be viewed as $\mathbf{Z}/n\mathbf{Z}$-modules. Moreover

$$\mathrm{Hom}(X, Y) = \mathrm{Hom}_{\mathbf{Z}/n\mathbf{Z}}(X, Y) \overset{\Psi}{\simeq} Y \qquad \text{where} \quad \Psi f = f(1),$$

and

$$W \otimes_{\mathbf{Z}} X = W \otimes_{\mathbf{Z}/n\mathbf{Z}} X \overset{\varphi}{\simeq} W \qquad \text{where} \quad \varphi(w \otimes x) = xw,$$

and the equality signs are the obvious identifications. Thus $h_w(x) = h_{xw}(1)$, and $\tilde{\beta}h$ is the composite isomorphism

$$W \otimes X \overset{\varphi}{\simeq} W \overset{h}{\simeq} \mathrm{Hom}(X, Y) \overset{\Psi}{\simeq} Y.$$

48.2 *Let G be a finite group. If* $H^{\iota}(G, \mathbf{Z})$ *is cyclic of order* $n = |G|$, *then so is* $H^{-\iota}(G, \mathbf{Z})$. *Moreover, if* ι_ι *is a generator of* $H^{\iota}(G, \mathbf{Z})$, *there exists a generator* $\iota_{-\iota}$ *of* $H^{-\iota}(G, \mathbf{Z})$ *such that* $\iota_{-\iota}\cdot\iota_\iota = 1$ *where the product is in the image of*

$$\mathrm{eval}^* \cup\colon H^{-\iota}(G, \mathrm{Hom}(\mathbf{Z}, \mathbf{Z})) \otimes H^{\iota}(G, \mathbf{Z}) \to H^0(G, \mathbf{Z}) = \mathbf{Z}/n\mathbf{Z}.$$

PROOF By 48.1, the isomorphism (12) of 47.10, in view of the identification

$$\mathrm{Hom}(H^{\iota}(G, \mathbf{Z}), \mathbf{Q}/\mathbf{Z}) \cong \mathrm{Hom}(H^{\iota}(G, \mathbf{Z}), \mathbf{Z}/n\mathbf{Z})$$

corresponds to the isomorphism

$$\tilde{\beta}\alpha = \mathrm{eval}^* \cup\colon H^{-\iota}(G, \mathbf{Z}) \otimes H^{\iota}(G, \mathbf{Z}) \simeq \mathbf{Z}/n\mathbf{Z}.$$

where we have identified $\mathbf{Z}$ with $\mathrm{Hom}(\mathbf{Z}, \mathbf{Z})$. Hence $H^{-l}(G, \mathbf{Z}) \cong \mathbf{Z}/n\mathbf{Z}$. Let ι_l be a generator of $H^l(G, \mathbf{Z})$; then there exists an element $\iota_{-l} \in H^{-l}(G, \mathbf{Z})$ such that

$$(\mathrm{eval}^* \cup)(i_{-l} \otimes \iota_l) = \iota_{-l} \cdot \iota_l = 1 \in \mathbf{Z}/n\mathbf{Z}.$$

Finally, the identifications above imply that ι_{-l} is a generator.

48.3 An integer l is called a *cohomological period* of the group G if $H^l(G, \mathbf{Z})$ is cyclic of order $|G|$.

48.4 (Artin and Tate [5]) *Let l be an even integer. Then l is a cohomological period of the finite group G if and only if*

$$H^r(G, C) \cong H^{r+l}(G, C) \qquad \text{for all } r \text{ and all } C. \tag{13}$$

PROOF *Necessity.* By 48.2, $H^{-l}(G, \mathbf{Z}) \cong \mathbf{Z}/n\mathbf{Z}$. Let ι_l and ι_{-l} be as in 48.2, i.e., $\iota_{-l} \cdot \iota_l = 1$. Define

$$\Phi: H^r(G, C) \xrightarrow{\sim} H^r(G, C) \otimes H^l(G, \mathbf{Z})$$

by $\Phi(a) = a \otimes \iota_l$. Identify C with $\mathrm{Hom}(\mathbf{Z}, C)$ and define

$$\gamma: H^r(G, C) \otimes H^l(G, \mathbf{Z}) \to H^{r+l}(G, C),$$

as in 47.2, by $\gamma(a \otimes \beta) = a \cdot \beta$. Define

$$\Phi^l: H^{r+l}(G, C) \xrightarrow{\sim} H^{r+l}(G, C) \otimes H^{-l}(G, \mathbf{Z})$$

by $\Phi'(a') = a' \otimes \iota_{-l}$. Finally identify C with $\mathrm{Hom}(\mathbf{Z}, C)$ and define

$$\gamma': H^{r+l}(G, C) \otimes H^{-l}(G, \mathbf{Z}) \to H^r(G, C),$$

as in 47.2, by $\gamma'(a' \otimes \beta') = a' \cdot \beta'$. Then

$$\begin{aligned} ((\gamma'\Phi')(\gamma\Phi))a &= (a \cdot \iota_l) \cdot \iota_{-l} \\ &= a \cdot (\iota_l \cdot \iota_{-l}) && \text{by 46.2} \\ &= (-1)^{-l^2} a \cdot (\iota_{-l} \cdot \iota_l) && \text{by 46.3} \\ &= a && \text{since } l \text{ is even,} \end{aligned}$$

and

$$
\begin{aligned}
((\gamma\Phi)(\gamma'\Phi'))a' &= (a\cdot\iota_{-l})\cdot\iota_l \\
&= a\cdot(\iota_{-l}\cdot\iota_l) \qquad \text{by 46.2} \\
&= a.
\end{aligned}
$$

Thus $\gamma\Phi: H^r(G, C) \to H^{r+l}(G, C)$ is an isomorphism.

Sufficiency. When $r = 0$ and $C = \mathbf{Z}$, (13) reduces to $H^0(G, \mathbf{Z}) \cong H^l(G, \mathbf{Z})$. Since $H^0(G, \mathbf{Z})$ is cyclic of order n (24 (vii)), so is $H^l(G, \mathbf{Z})$.

REMARK *If $G \neq \{1\}$, and l is a cohomological period of G, then l is even.* For, suppose l is odd and $H^l(G, \mathbf{Z})$ is cyclic of order $|G|$ generated by ι_l. Let ι_{-l} be the corresponding generator of $H^{-l}(G, \mathbf{Z})$. Then

$$\iota_l = \iota_l\cdot\iota_{-l}\cdot\iota_l = (-1)^{-l^2}\iota_{-l}\cdot\iota_l\cdot\iota_l = -\iota_l.$$

Then $2\iota_l = 0$ and so $|G| = 2$. But we know that $\mathbf{Z}/2\mathbf{Z}$ has even cohomological periods by Sections 23 and 24.

48.5 *If G has cohomological period l, then so does every subgroup U of G. Moreover, if ι_l is a generator of $H^l(G/\mathbf{Z})$, then so is $\operatorname{res}_{G,U}\iota_l \in H^l(U, \mathbf{Z})$.*

PROOF By 32.6, $\operatorname{cor}_{U,G} \operatorname{res}_{G,U} \iota_l = (G: U)\iota_l$. Since $(G: U)\iota_l$ has order $|U|$, the order of $\operatorname{res}_{G,U} \iota_l$ is at least $|U|$. But since no element of $H^l(U, \mathbf{Z})$ has order greater than $|U|$, we conclude that $\operatorname{res}_{G,U} \iota_l$ has order $|U|$, and thus is a generator of $H^l(U, \mathbf{Z})$.

CHAPTER

8

Spectral Sequences

In this chapter the group G is not necessarily finite and so, by Section 22, $H^0(G, M) = M^G$(and $H^n(G, M)$ makes sense only for $n \geqq 0$.

49. Differential Groups

Let A be an abelian group. An endomorphism $d_A: A \to A$ is called a *differential operator* if $d_A d_A = 0$. The abelian group A equipped with a differential operator is called a *differential group*. With differential groups as objects and homomorphisms which commute with d we get a category $\mathscr{G}_d$ called the category of differential groups. The group $H(A) = \ker d/\mathrm{im}\, d$ is called the *derived group* of A. If $f: A \to B$ is a morphism in $\mathscr{G}_d$, then $H(f): H(A) \to H(B)$ mapping $a + \mathrm{im}\, d_A \mapsto f(a) + \mathrm{im}\, d_B$ is well defined, for,

$$\begin{aligned} H(f)(a + d_A a' + \mathrm{im}\, d_A) &= f(a) + f d_A a' + \mathrm{im}\, d_B \\ &= f(a) + d_B f a' + \mathrm{im}\, d_B \\ &= f(a) + \mathrm{im}\, d_B. \end{aligned}$$

Moreover H is a functor from the category $\mathcal{G}_d$ to the category of abelian groups.

We say the morphisms $f, g: A \to B$ in $\mathcal{G}_d$ are *homotopic* and we write $f \sim g$ if there exists a morphism $\xi: A \to B$ such that

$$f - g = d_B \xi + \xi d_A.$$

49.1 *If $f \sim g$, then $H(f) = H(g)$.*

Let $0 \to A \xrightarrow{f} B \xrightarrow{g} C \to 0$ be an exact sequence in $\mathcal{G}_d$. Define the *connecting homomorphism*

$$H(C) \to H(A)$$

as follows: Let $c \in \ker d_C$. Choose $b \in B$ such that $gb = c$. Then $gd_Bb = d_Cgb = d_Cc = 0$. Hence there exists $a \in A$ such that $fa = d_Bb$; $a \in \ker d_A$, for $fd_Aa = d_Bfa = d_Bd_Bb = 0$ and f is a monomorphism. Define $\partial(c + \operatorname{im} d_C) = a + \operatorname{im} d_A$; ∂ is well defined, for, if $b' \in B$ is such that $g(b') = c$, then there exists $a'' \in A$ such that $fa'' = b - b'$. Let a' be obtained from b' in the same manner as a was obtained from b, then $d_A(a'') = a - a'$. Hence a and a' are representatives of the same coset $a + \operatorname{im} d_A$. Finally, using similar arguments one can show that $a + \operatorname{im} d_A$ is independent of the choice of the coset representative c. We point out that the construction of ∂ above is a special case of the homomorphism $\bar{d}_r$ in Section 8.3.

49.2 *If the sequence $0 \to A \xrightarrow{f} B \xrightarrow{g} C \to 0$ is exact in $\mathcal{G}_d$, then the triangle*

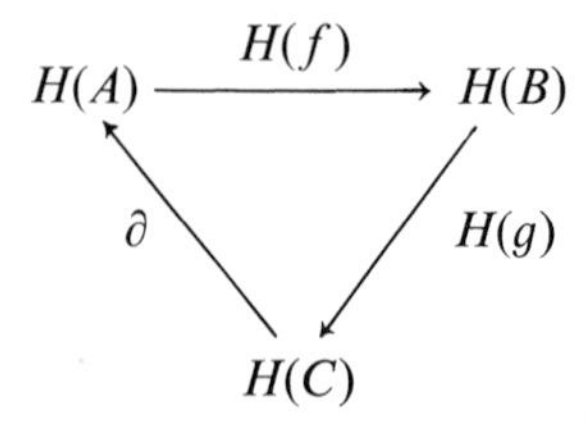

is exact in the category of abelian groups.

PROOF   Similar to the proof of 8.4.

49.3 *Graded and bigraded groups.* An abelian group A is called *graded* if for each integer n there is a subgroup A_n of A such that $A = \sum_n A_n$. The elements of A_n are called *homogeneous of degree n*. An abelian group A is called *bigraded* if for each pair of integers (p, q) there is a subgroup $A_{p,q}$ of A such that $A = \sum_{p,q} A_{p,q}$. The elements of $A_{p,q}$ are homogeneous of bidegree (p, q). Let A and B be bigraded groups. The homomorphism $f: A \to B$ is *homogeneous of bidegree* (p, q) if

$$f|_{A_{m,n}}: A_{m,n} \to A_{m+p,n+q}$$

for every pair (m, n).

49.4 *The spectral sequence of an exact couple* (Massey [71). An *exact couple* is a quintuple $\langle D_1, E_1, i_1, j_1, k_1\rangle$ where D_1 and E_1 are abelian groups, i_1, j_1, k_1 are homomorphisms, and the triangle

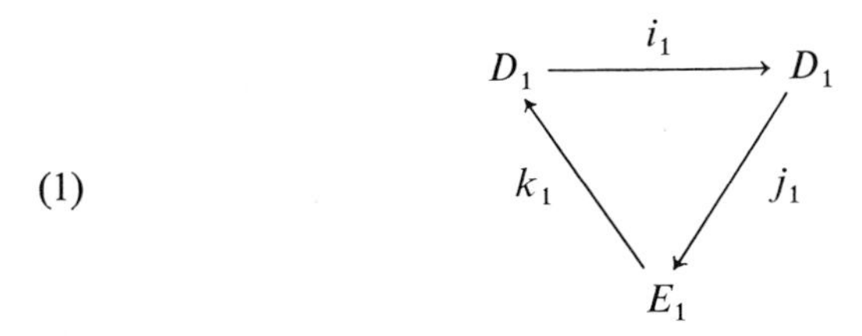

is exact. Define $d_1 = j_1k_1$. Then $d_1d_1 = 0$. Thus E_1 equipped with the endomorphism d_1 is a differential group. Let

(i) $E_2 = H(E_1)$ be the derived group of E_1

(ii) $D_2 = i_1(D_1)$,

(iii) $i_2 = i_1|_{D_2}$

(iv) $j_2: D_2 \to E_2$ be defined as follows: Let $x \in D_2$. Choose $y \in D_1$ such that $i_1y = x$. Then $j_1y \in \ker d_1$. We contend j_1y is uniquely determined modulo im d_1 by x. Namely, if $y' \in D_2$ is also such that $iy' = x$, then $i(y - y') = 0$; hence by the exactness of the triangle (1) there exists $z \in E_1$ such that $y - y' = k_1z$, so $j_1y - j_1y' = jk_1z = d_1z$. Define $j_2(x) = j_1y + \text{im } d_1$.

(v) $k_2: E_2 \to D_2$ be defined as follows: Since $k_1(\ker d_1) \subset D_2$ and $k_1(\text{im } d_1) = 0$, k_1 induces k_2. Explicitly $k_2(z + \text{im } d_1) = k_1z$ for every $z \in \ker d_1$. Then it is easy to verify that the triangle

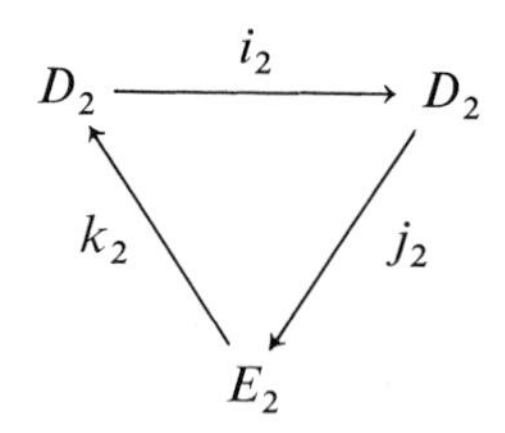

is exact. Hence the quintuple $\langle D_2, E_2, i_2, j_2, k_2\rangle$ is an exact couple. Continuing this process we obtain the sequence of differential groups $E_1, E_2, E_3, \ldots$, called the Koczul–Leray *spectral sequence* of the exact couple $\langle D_1, E_1, i_1, j_1, k_1\rangle$.

There is a canonical epimorphism h_n: $\ker d_n \to E_{n+1}$. Define

$$h_n^p = h_{n+p-1}h_{n+p-2}\cdots h_n.$$

So $h_n^1 = h_n$, and the domain of h_n^p for $p > 1$ is

$$Z_n^p = \{z \in Z_n^{p-1}, \text{ s.t. } h_n^{p-1}(z) \in \ker d_{n+p-1}\}.$$

Let $\bar{E}_n = \bigcap_p Z_n^p$. Define $\bar{h}_n = h_n|_{\bar{E}_n}: \bar{E}_n \to \bar{E}_{n+1}$. Then $\{\bar{E}_n, \bar{h}_n\}$ is a direct system of groups. The limit E_∞ of this direct system is called the limit group of the spectral sequence $E_1, E_2, E_3, \ldots$ (see Section 10).

In particular, if there exists an integer $r \geqq 1$ such that $d_n = 0$ for each $n \geqq r$, then $\ker d_n = E_n$, h_n: $E_n \cong E_{n+1}$, so $\bar{E}_n = E_n$ and $\bar{h}_n = h_n$ for each $n \geqq r$. Therefore in this case $E_\infty \cong E_r$.

Exercise An alternate definition of E_∞. Let $i_{(n)} = i_n i_{n-1}\cdots i_1$: $D_1 \to D_{n+1} \subset D_1$. Then $i_{(n)}$ is the n-fold iteration of i_1 since $i_n = i_1|_{D_n}$. Let

$$D_\infty = \bigcap_{n=1}^{\infty} D_n = \bigcap_{n=1}^{\infty} \operatorname{im} i_{(n)},$$

and $D_0 = \bigcup_{n=1}^{\infty} \ker i_{(n)}$. Show $E_\infty \cong k^{-1}(D_\infty)/j_1 D_0$.

49.5 *The exact couple associated with a filtration of a differential group.* A *filtration* of a differential group A is a family of subgroups $\{F^pA\}$ indexed by the integers such that $d(F^pA) \subset F^pA$ and

$$\cdots \supset F^pA \supset F^{p+1}A \supset \cdots.$$

For each index p the differential operator d induces the structure of a differential group on $F^pA/F^{p+1}A$. Hence for each p, the exactness of $0 \to F^{p+1}A \to F^pA \to F^pA/F^{p+1}A \to 0$, by 49.2, implies the exactness of the triangle

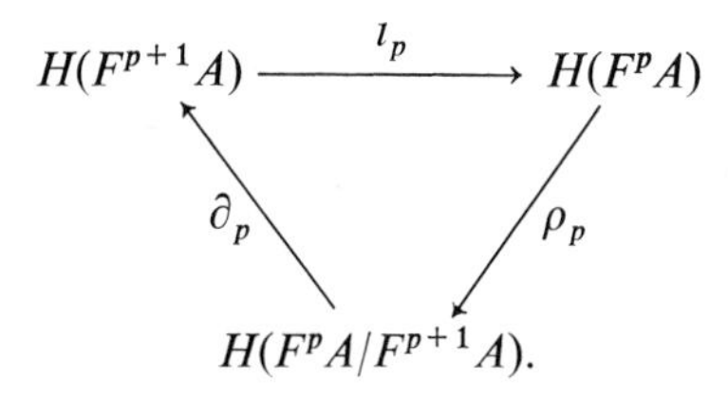

Let $D_1 = \sum_p H(F^pA)$, $E_1 = \sum_p H(F^pA/F^{p+1}A)$ where the sums are direct. Then the triangle

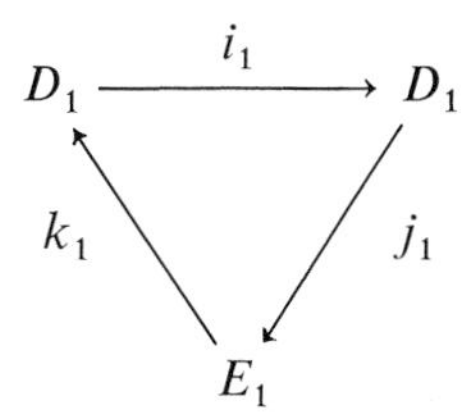

is exact where the maps i_1, j_1, k_1 are induced by the families of maps $\{\imath_p\}$, $\{\rho_p\}$, and $\{\partial_p\}$, respectively. Hence $\langle D_1, E_1, i_1, j_1, k_1 \rangle$ is an exact couple. The groups D_1, E_1 are graded and the maps i_1, j_1, k_1 are of degrees -1, 0, and 1, respectively.

49.6 *The exact couple associated with a filtration of a cochain complex.* Let

$$\cdots \to A^{p+q} \to A^{p+q+1} \to A^{p+q+2} \to \cdots$$

be a cochain complex with a filtration

$$\begin{array}{lcccccccc}
 & & \vdots & & \vdots & & \vdots & & \\
 & & \cup & & \cup & & \cup & & \\
(F^{p-1}A)^*: & \cdots \to & (F^{p-1}A)^{p+q} & \to & (F^{p-1}A)^{p+q+1} & \to & (F^{p-1}A)^{p+q+2} & \to & \cdots \\
 & & \cup & & \cup & & \cup & & \\
(F^pA)^*: & \cdots \to & (F^pA)^{p+q} & \to & (F^pA)^{p+q+1} & \to & (F^pA)^{p+q+2} & \to & \cdots \\
 & & \cup & & \cup & & \cup & & \\
(F^{p+1}A)^*: & \cdots \to & (F^{p+1}A)^{p+q} & \to & (F^{p+1}A)^{p+q+1} & \to & (F^{p+1}A)^{p+q+2} & \to & \cdots \\
 & & \cup & & \cup & & \cup & & \\
 & & \vdots & & \vdots & & \vdots & &
\end{array}$$

where the horizontal arrows are restrictions of the maps in the complex. Introduce a bigrading $F^{p,q}A = (F^pA)^{p+q}$. If

$$E_0^pA = \frac{F^pA}{F^{p+1}A},$$

then E_0^pA is graded and $E_0^{p,q} = (E_0^pA)^{p+q}$ furnishes it with a bigrading. The exactness of the sequence

$$0 \to (F^{p+1}A)^* \to (F^pA)^* \to (E_0^pA)^* \to 0,$$

of complexes by Section 8.4, gives rise to the long exact sequence

(2)

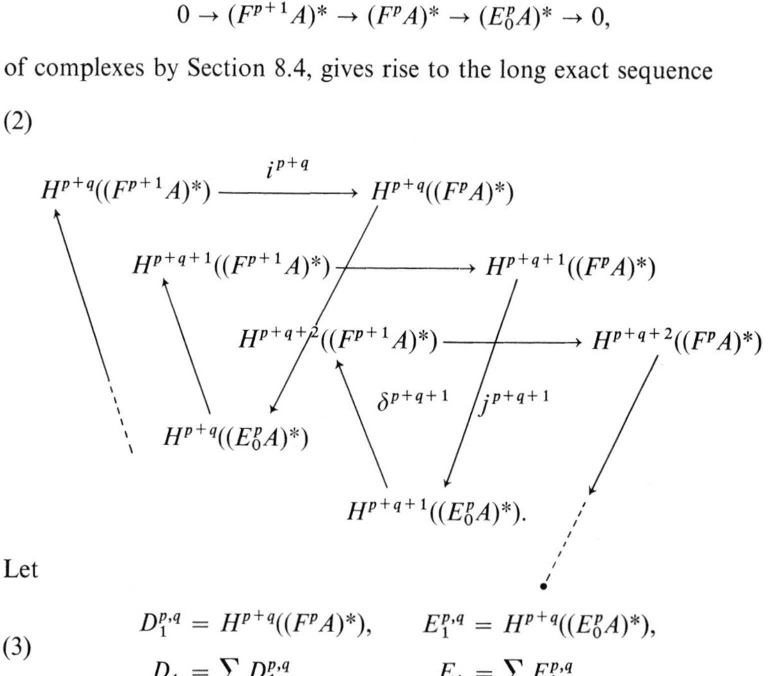

Let

(3)
$$D_1^{p,q} = H^{p+q}((F^pA)^*), \qquad E_1^{p,q} = H^{p+q}((E_0^pA)^*),$$
$$D_1 = \sum_{p,q} D_1^{p,q}, \qquad E_1 = \sum_{p,q} E_1^{p,q},$$

where the sums are direct. Then the exactness of the long sequence (2) implies the exactness of the diagram

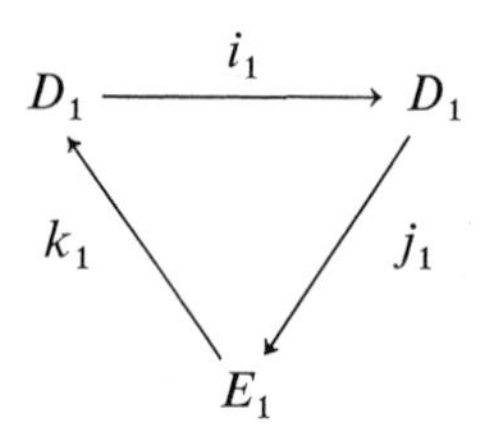

where i_1, j_1, k_1 are induced by the families of maps $\{i^{p+q}\}$, $\{j^{p+q}\}$, and $\{\delta^{p+q}\}$, respectively. Hence $\langle D_1, E_1, i_1, j_1, k_1\rangle$ is an exact couple. In view of the definitions (3), D_1 and E_1 are bigraded and:

i_1 is of bidegree $(-1, 1)$ since its restriction is

$$i_1^{p,q}: D_1^{p+1,q-1} \to D_1^{p,q},$$

j_1 is of bidegree $(0, 0)$ since its restriction is

$$j_1^{p,q}: D_1^{p,q} \to E_1^{p,q},$$

k_1 is of bidegree $(1, 0)$ since its restriction is

$$k_1^{p,q}: E_1^{p,q} \to D_1^{p+1,q}.$$

E_1 is a differential bigraded group and its differential operator is $d_1 = j_1k_1$; d_1 is of bidegree $(1, 0)$ since its restriction to $E_1^{p,q}$ is

$$d_1^{p,q} = j_1^{p+1,q} \cdot k_1^{p,q}: E_1^{p,q} \to E_1^{p+1,q}.$$

49.7 *The spectral sequence* (49.6 continued). Before the computation of the bidegree attached to $d_n^{p,q}$ we show here that the maps $d_1^{p,q}$ are the connecting homomorphisms δ^{p+q} in the long exact sequence obtained from the short exact sequence of complexes

$$0 \to (E_0^{p+1}A)^* \to (F^pA/F^{p+2}A)^* \to (E_0^pA)^* \to 0.$$

By Section 8.3,

$$\delta^{p+q}: H^{p+q}((F^pA/F^{p+1}A)^*) \to H^{p+q+1}((F^{p+1}A/F^{p+2}A)^*).$$

Moreover, from the commutativity of the diagram

$$\begin{array}{ccccccccc}
0 \to & (F^{p+1}A)^* & \to & (F^pA)^* & \to & (E_0^pA)^* & \to 0 \\
& \downarrow & & \downarrow & & \downarrow{\scriptstyle 1} & \\
0 \to & (E_0^{p+1}A)^* & \to & (F^pA/F^{p+2}A)^* & \to & (E_0^pA)^* & \to 0
\end{array}$$

where the rows are exact sequences of complexes, by Section 8.5, we get the commutative diagram

$$\begin{array}{ccc} H^{p+q}((E_0^p A)^*) & \xrightarrow{k_1^{p,q}} & H^{p+q+1}((F^{p+1}A)^*) \\ \Big\downarrow 1 & & \Big\downarrow j_1^{p+1,q} \\ H^{p+q}((E_0^p A)^*) & \xrightarrow{\delta^{p+q}} & H^{p+q+1}((E_0^{p+1}A)^*) \end{array}$$

where $k_1^{p,q}$ and δ^{p+q} are both connecting homomorphisms. Hence

$$\delta^{p+q} = d_1^{p,q}.$$

Using the constructions of D_2 and i_2 in 49.4,

$$D_2^{p,q} = \operatorname{im}(H^{p+q}((F^{p+1}A)^*) \xrightarrow{i_1^{p,q}} H^{p+q}((F^p A)^*)))$$

and

$$i_2^{p,q} = i_1^{p-1,q}|_{\operatorname{im} i_1{}^{p,q}}$$

This together with the constructions of E_2, j_2, and k_2 in 49.4 implies

$i_2^{p,q}$ is of bidegree $(-1, 1)$,
$j_2^{p,q}$ is of bidegree $(1, -1)$,
$k_2^{p,q}$ is of bidegree $(1, 0)$.

Repeating this process, for $n \geqq 2$

$$D_n^{p,q} = \operatorname{im}(H^{p+q}((F^{p+n-1}A)^*) \to H^{p+q}((F^p A)^*)))$$

and

$i_n^{p,q}$ is of bidegree $(-1, 1)$,
$j_n^{p,q}$ is of bidegree $(n - 1, -n + 1)$,
$k_n^{p,q}$ is of bidegree $(1, 0)$.

Hence we have the important result that *for $n \geqq 1$ the differential group E_n is bigraded and its differential operator d_n is of bidegree $(n, -n + 1)$. Moreover, the sequence*

$$D_n^{p-n+2,q+n-2} \xrightarrow{i_n} D_n^{p-n+1,q+n-1} \xrightarrow{j_n} E_n^{p,q} \xrightarrow{k_n} D_n^{p+1,q} \tag{4}$$

is exact.

N.B. In the sequel we will assume that the filtration of the complex is such that for each m and large enough n we have $(F^{p-n+2}A)^m = A^m$ and $(F^{p+n}A)^m = 0$. Under these assumptions the sequence (4) reduces, for large enough n, to the exact sequence

$$\begin{array}{c} \mathrm{im}(H^{p+q}((F^{p+1}A)^*) \to H^{p+q}(A^*)) \\ \Big\downarrow i_n \\ \mathrm{im}(H^{p+q}((F^{p}A)^*) \to H^{p+q}(A^*)) \\ \Big\downarrow j_n \\ E_n^{p,q} \\ \downarrow \\ 0 \end{array}$$

This means that for n large enough $E_n^{p,q}$ is independent of n, so $E_n^{p,q} = E_\infty^{p,q}$; moreover, if

$$F^pH^{p+q}(A^*) = \mathrm{im}(H^{p+q}((F^pA)^*) \to H^{p+q}(A^*)),$$

then we have the exact sequence

$$0 \to F^{p+1}H^{p+q}(A^*) \to F^pH^{p+q}(A^*) \to E_\infty^{p,q} \to 0.$$

Let $E_0^{p,q}H(A^*) = F^pH^{p+q}(A^*)/F^{p+1}H^{p+q}(A^*)$. Then

$$E_0H^{p+q}(A^*) = \sum_{i+j=p+q} E_0^{i,j}H(A^*)$$

is called *the bigraded group associated with* $H^{p+q}(A^*)$. Summarizing, we have shown that *if the filtration F of the complex A^* is such that for large enough n, $(F^{-n}A)^* = A^*$ and $(F^nA)^* = 0$, then $E_\infty^{p,q} \cong E_0^{p,q}H(A^*)$*. So for large enough n, $E_n^{p,q}$ is a subquotient of $H^{p+q}(A^*)$; and so, for instance, if all such sub quotients are zero, then $H^{p+q}(A^*)$ is zero.

50. How Spectral Sequences Are Used

50.1 Let A^*, A'^* be cochain complexes and $f\colon A^* \to A'^*$ a morphism of complexes. Suppose we can find filtrations F and F' of A^* and A'^* such that for each m there exists n such that $F^nA^m = F'^nA'^m = 0$ and $F^0A^* = A^*$, $F^0A'^* = A'^*$; moreover, the filtrations are compatible with the morphism f. If we want to show $H^*f\colon H^*(A^*) \simeq H^*(A'^*)$, we can use the following test: Construct the map $E_nf\colon E_nA^* \to E_nA'^*$. This induces the map $E_\infty f\colon E_\infty A^* \to E_\infty A'^*$. If $E_\infty f$ is an isomorphism, then H^*f is an isomorphism. Namely, suppose we have the isomorphism $E_0^*H(f)\colon E_0^* \simeq E_0^*H(A'^*)$. Then by the application of the Five Lemma to

$$\begin{array}{ccccccccc} 0 \to & F^nH^p(A^*) & \to & F^{n-1}H^p(A^*) & \to & F^{n-1}H^p(A^*)/F^nH^p(A^*) & \to 0 \\ & \downarrow\wr & & \downarrow & & \downarrow\wr & \\ 0 \to & F'^nH^p(A'^*) & \to & F'^{n-1}H^p(A'^*) & \to & F'^{n-1}H^p(A'^*)/F'^nH^p(A'^*) & \to 0 \end{array}$$

we have $F^{n-1}H^p(A^*) \cong F'^{n-1}H^p(A'^*)$. Again by the Five Lemma applied to

$$\begin{array}{ccccccccc} 0 \to & F^{n-1}H^p(A^*) & \to & F^{n-2}H^p(A^*) & \to & F^{n-2}H^p(A^*)/F^{n-1}H^p(A^*) & \to 0 \\ & \downarrow\wr & & \downarrow & & \downarrow\wr & \\ 0 \to & F'^{n-1}H^p(A'^*) & \to & F'^{n-2}H^p(A'^*) & \to & F'^{n-2}H^p(A'^*)/F'^{n-1}H^p(A'^*) & \to 0 \end{array}$$

we have $F^{n-2}H^p(A^*) \cong F'^{n-2}H^p(A'^*)$. Continuing this process we get $F^0H^p(A^*) \cong F'^0H^p(A'^*)$.

50.2 A^* is a cochain complex and the filtration F is such that $F^pA^{p+q} = 0$ if $p < 0$, $F^pA^{p+q} = A^{p+q}$ if $q < 0$, and for some large p_0, $F^pA^{p+q} = 0$ if $p \geqq p_0$. Then we have $E_2^{p,q} = 0$ if $p < 0$ or $q < 0$, and $d_2^{p,q}$ is of

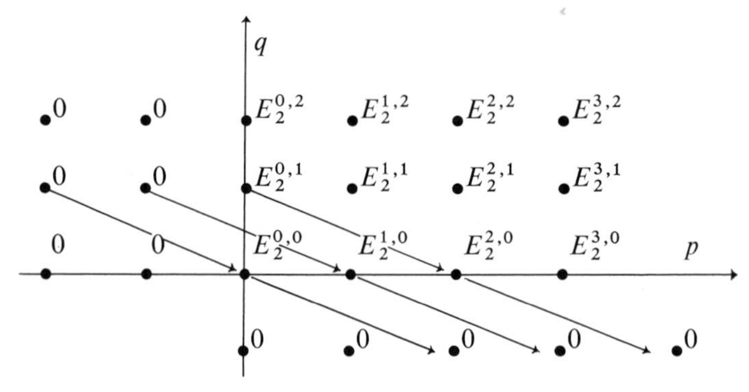

bidegree $(2, -1)$. Thus

$$E_2^{0,0} = E_\infty^{0,0} = H^0A^*.$$

We have $d_2^{1,0} = 0$, $d_2^{-1,1} = 0$, so

$$E_2^{1,0} = E_\infty^{1,0} = F^1H^1(A^*)/F^2H^1(A^*).$$

But $F^2H^1(A^*) = 0$ since $H^1(A^*) = F^0H^1(A^*) \supset F^1H^1(A^*) \supset 0$, so

$$E_2^{1,0} = F^1H^1(A^*). \tag{5}$$

Moreover $E_\infty^{0,1} = E_3^{0,1} = \ker(E_2^{0,1} \to E_2^{2,0})$, so

$$F^0H^1(A^*)/F^1H^1(A^*) = \ker(E_2^{0,1} \to E_2^{2,0}).$$

This together with the identity (5) implies $0 \to H^1(A^*)/E_2^{1,0} \to E_2^{0,1} \to E_2^{2,0}$ is exact. Splicing this with the exact sequence

$$0 \to E_2^{1,0} \to H^1(A^*) \to H^1(A^*)/E_2^{1,0} \to 0$$

we get the exact sequence

$$0 \to E_2^{1,0} \to H^1(A^*) \to E_2^{0,1} \to E_2^{2,0} \to H^2(A^*) \tag{6}$$

since $\operatorname{coker}(E_2^{0,1} \to E_2^{2,0}) = E_3^{2,0} = E_\infty^{2,0} = F^2H^2(A^*)$, coker meaning cokernel. Moreover if $E_2^{0,2} = 0$,

$$\ker(E_2^{1,1} \to E_2^{3,0}) = E_3^{1,1} = E_\infty^{1,1} = H^2(A^*)/F^2H^2(A^*),$$

hence

$$\begin{aligned} 0 \to E_2^{1,0} \to H^1(A^*) \to E_2^{0,1} &\xrightarrow{d_2^{0,1}} E_2^{2,0} \to H^2(A^*) \\ \to E_2^{1,1} &\xrightarrow{d_2^{1,1}} E_2^{3,0} \to H^3(A^*) \end{aligned} \tag{6'}$$

is exact.

50.3 If the filtration of the cochain complex is such that E_2 reduces to two rows

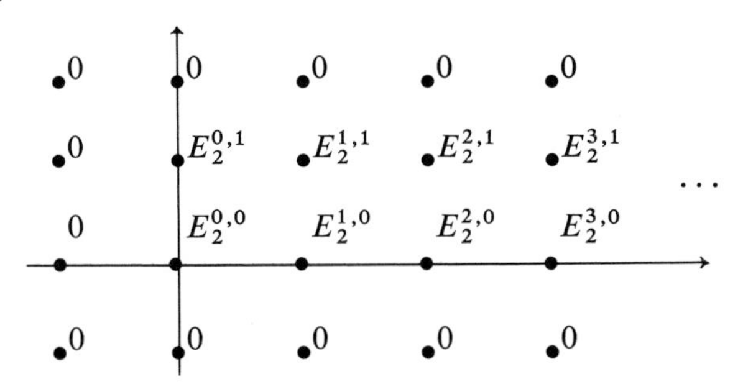

such that $E_2^{p,q} = 0$ if $q \neq 0, 1$, then $E_3^{p,q} = E_\infty^{p,q}$, and the exact sequence (6′) above can be extended to the exact sequence

$$0 \to E_2^{1,0} \to H^1(A^*) \to \cdots \to H^{p-1}(A^*) \to E_2^{p-2,1} \xrightarrow{d_2^{p-2,1}} E_2^{p,0} \to H^p(A^*) \to \cdots.$$

50.4 When E_2 reduces to $E_2^{p,q} = 0$ if $p \neq 0, 1$, then $E_2^{p,q} = E_\infty^{p,q}$ and we have the exact sequence

$$0 \to E_2^{0,q} \to H^q(A^*) \to E_2^{1,q-1} \to 0.$$

51. The Cartan–Leray Spectral Sequence

If a filtration of a cochain complex is given which is compatible with the grading, the resulting spectral sequence is not an invariant of the complex. However, for some filtrations the term $E_2^{p,q}$ is an invariant of the complex. In this section we compute the term $E_2^{p,q}$ of a spectral sequence first studied by Cartan and Leray [14, 16].

51.1 Let G be a group,

$$X_*: X_0 \xleftarrow{\partial_1} X_1 \xleftarrow{\partial_2} X_2 \longleftarrow \cdots$$

the normalized bar resolution of G, and

$$C^*: C^0 \xrightarrow{\delta^0} C^1 \xrightarrow{\delta^1} C^2 \longrightarrow \cdots$$

a cochain complex of G-modules. Let $A^{p,q} = \mathrm{Hom}_G(X_p, C^q)$ and $A = \sum_{p,q} A^{p,q}$. Thus A is a bigraded group on which we can define two differential operators as follows: $d_G: A \to A$ where

$$d_G|_{A^{p,q}} = \mathrm{Hom}_G(\partial_{p+1}, C^q): A^{p,q} \to A^{p+1,q}$$

and d_C $A \to A$ where $d_C|_{A^{p,q}}: A^{p,q} \to A^{p,q+1}$ is defined by $(d_C\varphi)[g_1, \ldots, g_p] = \delta^q(\varphi[g_1, \ldots, g_p])$ where δ^q is the qth coboundary homomorphism of the complex C^*, and $\varphi \in \mathrm{Hom}_G(X_p, C^q)$.

Since $C^q = \mathrm{Hom}_G(\mathbf{Z}G, C^q) = A^{0,q}$, we may view C^* as a subgroup of

A. The endomorphism $d = d_G + (-1)^p d_C$ is a differential operator and furnishes A with the structure of a differential group. The subgroup C^* of A is not invariant under the endomorphism d, but the subgroup $(C^*)^G = \{c \in C^* \text{ s.t. } gc = c\}$ is invariant under d, i.e., $d(C^*)^G \subset (C^*)^G$. For, if $c \in (C^q)^G$, then $dc = d_C c$. Hence $(C^*)^G$ may be viewed as a differential subgroup of the differential group A. Therefore the bigraded group A has the structure of a cochain complex

$$A^0 \xrightarrow{d} A^1 \xrightarrow{d} A^2 \rightarrow \cdots$$

where $A^n = \sum_{p+q=n} A^{p,q}$, and $(C^*)^G$ is a cochain subcomplex of A.

51.2 *If $H^n(G, C^q) = 0$ for all $q \geqq 0$ and all $n > 0$, then the inclusion of the cochain complex $(C^*)^G$ into the cochain complex A induces an isomorphism $H^n((C^*)^G) \cong H^n(A)$, for all $n \geqq 0$.*

PROOF Let

$$M^i = \sum_{q \geqq i} \sum_{p=0}^{\infty} A^{p,q}, \qquad N^i = M^i \cap (C^*)^G = \sum_{q \geqq i} (C^q)^G.$$

Then M^i and N^i equipped with the differential map d are cochain complexes; M^i/M^{i+1} is the cochain complex $\text{Hom}_G(X_*, C^i)$, and N^i/N^{i+1} is the cochain complex $(C^i)^G \xrightarrow{0} 0 \xrightarrow{0} \cdots$. Therefore the hypothesis that $H^n(G, C^i) = 0$ for all $n > 0$ implies that the inclusion of the complex N^i/N^{i+1} into the complex M^i/M^{i+1} induces an isomorphism

$$H^n(N^i/N^{i+1}) \cong H^n(M^i/M^{i+1}) \tag{7}$$

for every $n \geqq 0$. We contend this implies

$$H^n(N^0/N^p) \cong H^n(M^0/M^p) \tag{8}$$

for every $p \geqq 0$. Namely the diagram of complexes

$$\begin{array}{ccccccccc}
0 \rightarrow & N^1/N^2 & \rightarrow & N^0/N^2 & \rightarrow & N^0/N^1 & \rightarrow 0 \\
& \downarrow & & \downarrow & & \downarrow & \\
0 \rightarrow & M^1/M^2 & \rightarrow & M^0/M^2 & \rightarrow & M^0/M^1 & \rightarrow 0
\end{array}$$

by (7) and Section 8.5, implies the commutative diagram

$$\begin{array}{ccccccccc} H^{n-1}(N^0/N^1) & \to & H^n(N^1/N^2) & \to & H^n(N^0/N^2) & \to & H^n(N^0/N^1) & \to & H^{n+1}(N^1/N^2) \\ \downarrow\wr & & \downarrow\wr & & \downarrow\alpha & & \downarrow\wr & & \downarrow\wr \\ H^{n-1}(M^0/M^1) & \to & H^n(M^1/M^2) & \to & H^n(M^0/M^2) & \to & H^n(M^0/M^1) & \to & H^{n+1}(M^1/M^2) \end{array}$$

where the rows are exact. Hence, by the Five Lemma, α is an isomorphism. Next consider the diagram of complexes

$$\begin{array}{ccccccccc} 0 & \to & N^2/N^3 & \to & N^0/N^3 & \to & N^0/N^2 & \to & 0 \\ & & \downarrow & & \downarrow & & \downarrow & & \\ 0 & \to & M^2/M^3 & \to & M^0/M^3 & \to & M^0/M^2 & \to & 0. \end{array}$$

Again apply (7) and Section 8.5 and use the fact that α is an isomorphism to get $H^n(N^0/N^3) \cong H^n(M^0/M^3)$. Repeat this process to get (8). Then for fixed n, 51.2 follows by setting $p \geqq n + 2$ in (8).

51.3 *The spectral sequence.* Let A be the bigraded complex defined in 51.1. Let $F^iA = \sum_{p \geqq i} \sum_{q=0}^{\infty} A^{p,q}$. Then

$$A = F^0A \supset F^1A \supset \cdots, \qquad d(F^iA) \subset F^iA, \qquad \text{and} \qquad A^{p,q} \cap F^iA = 0$$

if $i > p$. Hence the groups F^iA define a filtration of A. The spectral sequence associated with this filtration is called the *Cartan–Leray spectral sequence*. By 49.6, $E_1^{p,q} = H^{p+q}(F^pA/F^{p+1}A)$, where the cochain complex $F^pA/F^{p+1}A$ with the coboundary map induced by d is the complex $\sum_{q=0}^{\infty} A^{p,q} = \mathrm{Hom}_G(X_p, C^*)$ with the coboundary map $(-1)^p\,\delta^q$. Hence the term $E_1^{p,q}$ of the spectral sequence is isomorphic with $\mathrm{Hom}_G(X_p, H^q(C^*))$.

To compute $E_2^{p,q}$ recall that in 49.7 we have shown that $d_1^{p,q}$ is the connecting homomorphism associated with the exact sequence of complexes

$$(9) \qquad 0 \to F^{p+1}A/F^{p+2}A \overset{\alpha}{\to} F^pA/F^{p+2}A \overset{\beta}{\to} F^pA/F^{p+1}A \to 0.$$

We claim that under the identification $E_1^{p,q} = \mathrm{Hom}_G(X_p, H^q(C^*))$, $d_1^{p,q} = \mathrm{Hom}_G(\partial^{p+1}, H^q(C^*))$. For, let $\bar{\varphi} \in \mathrm{Hom}_G(X_p, H^q(C^*))$ and $\varphi \in \mathrm{Hom}_G(X_p, C^q)$ a cocycle in the complex $\mathrm{Hom}_G(X_p, C^*) = F^pA/F^{p+1}A$ such that for any generator $[g_1, \ldots, g_p]$ of X_p, the cohomology class of $\varphi[g_1, \ldots, g_p]$ in C^* is $\bar{\varphi}[g_1, \ldots, g_p]$; φ considered as a cochain in $F^pA/F^{p+2}A$ is such that $\beta(\varphi) = \varphi$ where β is the epimorphism in (9). Then by the definition of connecting homomorphism

$$d\varphi = d_G\varphi + (-1)^p d_C\varphi = d_G\varphi \in \mathrm{Hom}_G(X_{p+1}, C^q)$$

is a cocycle in the complex $F^{p+1}A/F^{p+2}A = \mathrm{Hom}_G(X_{p+1}, C^*)$, and therefore it represents an element $\bar{\psi} \in E_1^{p+1,q} = \mathrm{Hom}_G(X_{p+1}, H^q(C^*))$. Thus $d_1^{p,q}\bar{\varphi} = \bar{\psi}$ and $d_1^{p,q} = \mathrm{Hom}_G(\partial^{p+1}, H^q(C^*))$. Summarizing, *the term $E_2^{p,q}$ of the spectral sequence associated with the filtration $\{F^iA\}$ is isomorphic with $H^p(G, H^q(C^*))$.*

As in every spectral sequence the group $E_\infty^{p,q}$ is isomorphic with the subquotient $F^pH^{p+q}(A)/F^{p+1}H^{p+q}(A)$ of $H^{p+q}(A)$ where the filtration $\{F^pH^{p+q}(A)\}$ is induced by the filtration $\{F^iA\}$ of A (see 49.7). Combining these results with 51.2 we have the following:

51.4 *Let C^* be a cochain complex of G-modules. Let X_* be the normalized bar resolution of G. Then in the spectral sequence $\{E_r\}$ associated with the filtration*

$$F^iA = \sum_{p \geqq i} \sum_{q=0}^{\infty} \mathrm{Hom}_G(X_p, C^q)$$

of the differential module

$$A = \sum_{p=0}^{\infty} \sum_{q=0}^{\infty} \mathrm{Hom}_G(X_p, C^q),$$

the term $E_2^{p,q}$ is isomorphic with $H^p(G, H^q(C^*))$. Moreover, if $H^n(G, C^q) = 0$ for $q \geqq 0$ and $n > 0$, then the term $E_\infty^{p,q}$ is isomorphic with the subquotient $F^pH^{p+q}((C^*)^G)/F^{p+1}H^{p+q}((C^*)^G)$ of $H^{p+q}((C^*)^G)$ where the filtration $\{F^pH^{p+q}((C^*)^G)\}$ is induced by the isomorphism $H^n((C^*)^G) \cong H^n(A)$.

52. The Spectral Sequence of a Group Extension

52.1 Let G be a group, N a normal subgroup of G, A a G-module, X_* the normalized bar resolution of G, and $\bar{X}_*$ the normalized bar resolution of G/N. Recall that $\mathrm{Hom}(X_*, A)$ is the cochain complex whose qth term is the abelian group of all **Z**-homomorphisms $\mathrm{Hom}_{\mathbf{Z}}(X_q, A)$, and $\mathrm{Hom}_N(X_*, A)$ is the cochain complex whose qth term is $\mathrm{Hom}_N(X_q, A) =$

$(\mathrm{Hom}_{\mathbf{Z}}(X_q, A))^N$. Define an action of G/N on $\mathrm{Hom}_N(X_q, A)$ by

$$\varphi^{\bar{g}}(g_0[g_1, \ldots, g_q]) = g(\varphi(g^{-1}g_0[g_1, \ldots, g_q]))$$

where $\bar{g} = g\,N \in G/N$, $\varphi \in \mathrm{Hom}_N(X_q, A)$, and $g_0[g_1, \ldots, g_q] \in X_q$. This action is well defined since φ is an N-homomorphism. Consider the cochain complex $\mathrm{Hom}_{G/N}(\overline{X}_*, \mathrm{Hom}_N(X_q, A))$. We contend that the homology groups of this complex, $H^p(G/N, \mathrm{Hom}_N(X_q, A))$, are zero for $p > 0$ and $q \geqq 0$. Namely, let $\varphi\colon \overline{X}_p \to \mathrm{Hom}_N(X_q, A)$ be a cocycle. Define $\psi\colon \overline{X}_{p-1} \to \mathrm{Hom}_N(X_q, A)$ by

$$\begin{aligned}(\psi(\bar{g}[\bar{g}_1, \ldots, \bar{g}_{p-1}]))&g'_0[g'_1, \ldots, g'_q] \\ &= (\varphi(\bar{g}'_0[\bar{g}'^{-1}_0\bar{g}, \bar{g}_1, \ldots, \bar{g}_{p-1}]))g'_0[g'_1, \ldots, g'_q].\end{aligned}$$

ψ is a G/N-homomorphism, for,

$$\begin{aligned}(\psi^{\bar{g}}&[\bar{g}_1, \ldots, \bar{g}_{p-1}])g'_0[g'_1, \ldots, g'_q] \\ &= (\bar{g}\cdot\psi(\bar{g}^{-1}[\bar{g}_1, \ldots, \bar{g}_{p-1}]))g'_0[g'_1, \ldots, g'_q] \\ &= \bar{g}((\psi(\bar{g}^{-1}[\bar{g}_1, \ldots, \bar{g}_{p-1}]))g^{-1}g'_0[g'_1, \ldots, g'_q]) \\ &= \bar{g}(\varphi(\bar{g}^{-1}\bar{g}'_0[\bar{g}'^{-1}_0\bar{g}\bar{g}^{-1}, \bar{g}_1, \ldots, \bar{g}_{p-1}])g^{-1}g'_0[g'_1, \ldots, g_q]) \\ &= (\varphi(\bar{g}^{-1}\bar{g}'_0[\bar{g}'^{-1}_0, \bar{g}_1, \ldots, \bar{g}_{p-1}]))^{\bar{g}}\bar{g}'_0[g'_1, \ldots, g'_q] \\ &= (\varphi(\bar{g}'_0[\bar{g}'^{-1}_0, \bar{g}_1, \ldots, \bar{g}_{p-1}]))g'_0[g'_1, \ldots, g'_q] \\ &= (\psi[\bar{g}_1, \ldots, \bar{g}_{p-1}])g'_0[g'_1, \ldots, g'_q].\end{aligned}$$

Moreover, it follows from the definitions and the fact that φ is a cocycle that

$$\begin{aligned}((\mathrm{Hom}_{G/N}(\partial_p, 1)\psi)&[\bar{g}_1, \ldots, \bar{g}_p])g'_0[g'_1, \ldots, g'_q] \\ &= (\varphi[\bar{g}_1, \ldots, \bar{g}_p])g'_0[g'_1, \ldots, g'_q].\end{aligned}$$

Thus $H^p(G/N, \mathrm{Hom}_N(X_q, A)) = 0$ for $p > 0$ and $q \geqq 0$. Hence we can apply 51.4 with G replaced by G/N, $C^* = \mathrm{Hom}_N(C_*, A)$. Then

$$(C^*)^{G/N} = (\mathrm{Hom}(C_*, A))^G = \mathrm{Hom}_G(C_*, A),$$

so that $H^n((C^*)^{G/N}) = H^n(G, A)$, and $H^n(C^*) = H^n(N, A)$. We have the following:

52.2 *Let G be a group, $N \trianglelefteq G$, A a G-module. Then there exists a* *spectral sequence* $\{E_r\}$ *of which the term* $E_2^{p,q}$ *is isomorphic with* $H^p(G/N, H^q(N, A))$, *and* $E_\infty^{p,q}$ *is isomorphic with some subquotient of* $H^{p+q}(G, A)$.

It is customary to report a statement of the above type partially by the notation

$$B^{p,q} \underset{p}{\Rightarrow} D^{p+q}.$$

This notation conveys the information that there is a complex A with a regular filtration such that the term $E_2^{p,q}$ is isomorphic with $B^{p,q}$, the degree of the filtration is p (under the arrow), and the term $E_\infty^{p,q}$ is isomorphic with a subquotient of D^{p+q}, and moreover, $H^{p+q}(A)$ is isomorphic with D^{p+q}. By "a *regular filtration*" in the sentence above we mean the filtration F of the complex A is such that for each n there exists an integer $m(n)$ such that $H^n(F^l(A)) = 0$ for $l > m(n)$.

Thus the result stated in 52.2 is in part

$$H^p(G/N, H^q(N, A)) \underset{p}{\Rightarrow} H^{p+q}(G, A).$$

53. The Hochschild–Serre Spectral Sequence

Let G be a group N, a normal subgroup of G. In this section we will obtain a spectral sequence (the Hochschild–Serre [51] spectral sequence) by a filtration of the cochain groups $\mathrm{Hom}_G(X_p, A)$ where A is a G-module and $X_*: X_0 \leftarrow X_1 \leftarrow \cdots$ is the normalized bar resolution of G. We will show that this filtration results in

$$H^p(G/N, H^q(N, A)) \underset{p}{\Rightarrow} H^{p+q}(G, A).$$

53.1 *The filtration.* Write $A^n = \mathrm{Hom}_G(X_n, A)$, and $A^* = \sum_{n=0}^{\infty} A^n$. Let $N \trianglelefteq G$, and $X_{N,*}$ be the normalized bar resolution of N. We define the filtration $\{F^pA^*\}$ as follows:

$$F^pA^* = A^* \qquad \text{for} \quad p \leqq 0,$$

and

$$F^pA^* = \sum_{n=0}^{\infty} F^pA^* \cap A^n \qquad \text{for} \quad p > 0,$$

where $F^pA^* \cap A^n = 0$ for $p > n$; and for $0 < p \leqq n$, $F^pA^* \cap A^n$ is the group of all $\varphi \in A^n$ for which $\varphi[g_1, \ldots, g_n]$ depends only on $g_1, \ldots, g_{n-p}$ and the cosets $g_{n-p+1}N, \ldots, g_nN$. Clearly $\delta(F^pA^*) \subset F^pA^*$.

To compute $E_1^{p,q}$ define the map of complexes

$$\rho_p: F^pA^* \to \operatorname{Hom}_{G/N}(\overline{X}_p, \operatorname{Hom}_N(X_{N,*}, A))$$

by

$$((\rho_p\varphi)[\bar{g}_1, \ldots, \bar{g}_p])[g'_1, \ldots, g'_q] = \varphi[g'_1, \ldots, g'_q, g_1, \ldots, g_p]$$

for $\varphi \in F^pA^* \cap A^{p+q}$ where $g_i \in G$ is a representative of the coset $\bar{g}_i$ and $g'_j \in N$. It is clear from the formula for coboundary homomorphism that ρ_p is a morphism of complexes, and

$$\rho_p(F^pA^* \cap A^{p+q}) \subset \operatorname{Hom}_{G/N}(\overline{X}_p, \operatorname{Hom}_N(X_{N,q}, A)).$$

Furthermore the restriction of ρ_p to the subcomplex $F^{p+1}A^*$ of F^pA^* is the trivial morphism of complexes. Therefore ρ_p induces a morphism of complexes

$$F^pA^*/F^{p+1}A^* \to \operatorname{Hom}_{G/N}(\overline{X}_p, \operatorname{Hom}_N(X_{N,*}, A)),$$

hence, a homomorphism $\Phi^{p,q}$ of $E_1^{p,q} = H^{p+q}(F^pA^*/F^{p+1}A^*)$ into $\operatorname{Hom}_{G/N}(\overline{X}_p, H^q(N, A))$.

53.2 $E_1^{p,q} \overset{\Phi^{p,q}}{\cong} \operatorname{Hom}_{G/N}(\overline{X}_p, H^q(N, A))$ *and this isomorphism is induced by the morphism* $\rho_p: F^pA^* \to \operatorname{Hom}_{G/N}(\overline{X}_p, \operatorname{Hom}_N(X_{N,*}, A))$.

Proof To show $\Phi^{p,q}$ is a monomorphism, let $\varphi \in F^pA^* \cap A^{p+q+1}$ such that $\delta\varphi \in F^{p+1}A^* \cap A^{p+q+2}$, i.e., φ represents a $p + q + 1$-cocycle in $F^pA^*/F^{p+1}A^*$. Suppose $\rho_p\varphi = \delta u$ for some $u \in \operatorname{Hom}_{G/N}(\overline{X}_p, \operatorname{Hom}_N(X_{N,q}, A))$. We must show there exists $v \in F^pA^* \cap A^{p+q}$ such that $\varphi - \delta v \in F^{p+1}A^* \cap A^{p+q+1}$.

To construct v define, for $g'_1, \ldots, g'_q$ in N and $g_1, \ldots, g_p$ in G, the map

$$\gamma: X_{N,q} \times X_p \to A$$

by

$$\gamma(g'_1, \ldots, g'_q, g_1, \ldots, g_p) = (u[\bar{g}_1, \ldots, \bar{g}_p])[g'_1, \ldots, g'_q]$$

where $\bar{g}_i = g_i N$. If $q = 0$ (which is the case $q = 1$ in the statement of 53.2), from $(\delta\varphi)(g, g', g_1, \ldots, g_p) = 0$, for $g' \in N$ we obtain

$$\begin{aligned}\varphi(gg', g_1, \ldots, g_p) &= g\varphi(g', g_1, \ldots, g_p) + \varphi(g, g'g_1, g_2, \ldots, g_p)\\ &= g((\rho_p\varphi)[\bar{g}_1, \ldots, \bar{g}_p])[g'] + \varphi(g, g_1, g_2, \ldots, g_p)\\ &= gg'\cdot\gamma(g_1, \ldots, g_p) - g\cdot\gamma(g_1, \ldots, g_p)\\ &\quad + \varphi(g, g_1, \ldots, g_p),\end{aligned}$$

which differs from $\delta\gamma(gg', g_1, \ldots, g_p)$ by terms that are independent of $g' \in N$. Hence $(\varphi - \delta\gamma)(gg', g_1, \ldots, g_p)$ is independent of g', so $\varphi - \delta\gamma \in F^{p+1}A^* \cap A^{p+1}$. Therefore if $q = 0$, we may choose $v = \gamma$.

If $q > 0$, construct the sequence $\gamma_0 = \gamma, \gamma_1, \ldots, \gamma_q$ as follows:

$$\gamma_k: X_k \times X_{N,q-k} \times X_p \to A$$

is defined recursively by the formulas

$$\begin{aligned}&\gamma_1(gg', g'_1, \ldots, g'_{q-1}, g_1, \ldots, g_p)\\ &\quad = g\gamma(g', g'_1, \ldots, g'_{q-1}, g_1, \ldots, g_p) - \varphi(g, g', g'_1, \ldots, g'_{q-1}, g_1, \ldots, g_p)\end{aligned}$$

and for $k > 1$

$$\begin{aligned}&\gamma_k(g^{(1)}, \ldots, g^{(k-1)}, gg', g'_1, \ldots, g'_{q-k}, g_1, \ldots, g_p)\\ &\quad = \gamma_{k-1}(g^{(1)}, \ldots, g^{(k-2)}, g^{(k-1)}g, g', g'_1, \ldots, g'_{q-k}, g_1, \ldots, g_p)\\ &\qquad + (-1)^k\varphi(g^{(1)}, \ldots, g^{(k-1)}, g, g', g'_1, \ldots, g'_{q-k}, g_1, \ldots, g_p).\end{aligned}$$

We observe that for $k \geqq 1$,

$$\begin{aligned}&\gamma_k(g^{(1)}, \ldots, g^{(k-1)}, g', g'_1, \ldots, g'_{q-k}, g_1, \ldots, g_p)\\ &\quad = \gamma_{k-1}(g^{(1)}, \ldots, g^{(k-1)}, g', g'_1, \ldots, g'_{q-k}, g_1, \ldots, g_p),\end{aligned}$$

so γ_k is an extension of γ_{k-1}. Furthermore

$$\begin{aligned}&\delta\gamma_k(g^{(1)}, \ldots, g^{(k-1)}, g', g'_1, \ldots, g'_{q+1-k}, g_1, \ldots, g_p)\\ &\quad = \delta\gamma_{k-1}(g^{(1)}, \ldots, g^{(k-1)}, g', g'_1, \ldots, g'_{q+1-k}, g_1, \ldots, g_p).\end{aligned}$$

It follows that

$$\gamma_k(g^{(1)}, \ldots, g^{(l-1)}, g, g'_1, \ldots, g'_{q-l}, g_1, \ldots, g_p)$$
$$= \gamma_l(g^{(1)}, \ldots, g^{(l-1)}, g, g'_1, \ldots, g'_{q-l}, g_1, \ldots, g_p) = 0$$

for $1 \leqq l \leqq k$. Hence by the coboundary formula

$$\begin{aligned}\delta\gamma_k(g^{(1)}, &\ldots, g^{(k-1)}, g, g', g'_1, \ldots, g'_{q-k}, g_1, \ldots, g_p)\\ &= (-1)^k\gamma_k(g^{(1)}, \ldots, g^{(k-1)}, gg', g'_1, \ldots, g'_{q-k}, g_1, \ldots, g_p)\\ &\quad + (-1)^{k-1}\gamma_k(g^{(1)}, \ldots, g^{(k-2)}, g^{(k-1)}g, g', g'_1, \ldots, g'_{q-k}, g_1, \ldots, g_p)\\ &= \varphi(g^{(1)}, \ldots, g^{(k-1)}, g, g', g'_1, \ldots, g'_{q-k}, g_1, \ldots, g_p)\end{aligned}$$

for $k > 1$, and

$$\begin{aligned}\delta\gamma_1(g, g'_1, \ldots, g'_q, g_1, \ldots, g_p) &= g\gamma_1(g'_1, \ldots, g'_q, g_1, \ldots, g_p)\\ &\quad - \gamma_1(gg'_1, g'_2, \ldots, g'_q, g_1, \ldots, g_p)\\ &= \varphi(g, g'_1, \ldots, g'_q, g_1, \ldots, g_p).\end{aligned}$$

Thus for all $k \geqq 1$,

(10) $$(\varphi - \delta\gamma_k)(g^{(1)}, \ldots, g^{(k-1)}, g, g', g'_1, \ldots, g'_{q-k}, g_1, \ldots, g_p) = 0.$$

We claim the same relation holds if g is replaced by gg'_0, where $g'_0 \in N$. Using induction on k, observe that $(\varphi - \delta\gamma_0)(g'_0, g'_1, \ldots, g'_q, g_1, \ldots, g_p) = 0$. Suppose we have the identity

$$(\varphi - \delta\gamma_{k-1})(g^{(1)}, \ldots, g^{(k-1)}, g'_0, g', g'_1, \ldots, g'_{q-k}, g_1, \ldots, g_p) = 0.$$

Expand

$$(\varphi - \delta\gamma_k)(g^{(1)}, g^{(k-1)}, g, g'_0, g', g'_1, \ldots, g'_{q-k}, g_1, \ldots, g_p) = 0$$

using the coboundary formula, and write the term

(11) $$(\varphi - \delta\gamma_k)(g^{(1)}, \ldots, g^{(k-1)}, gg'_0, g', g'_1, \ldots, g'_{q-k}, g_1, \ldots, g_p)$$

as a linear combination of the remaining terms with coefficients ± 1, which are values of $(\varphi - \delta\gamma_k)$ at $p + q + 1$-tuples in which the kth entries are either g or g'_0. The terms with g in the kth entry are zero by (10). The terms with g'_0 in the kth entry remain the same if γ_k is replaced by γ_{k-1}, and are zero by the inductive hypothesis. Hence for all $k \geqq 1$

the expression (11) is zero. Thus for $k = q$ we have

$$(\varphi - \delta\gamma_q)(g^{(1)}, \ldots, g^{(q)}, g'_0, g_1, \ldots, g_p) = 0.$$

Therefore if we write $(\varphi - \delta\gamma_q)(g^{(1)}, \ldots, g^{(q)}, gg'_0, g_1, \ldots, g_p)$ as above in terms of values of $\pm(\varphi - \delta\gamma_q)$ with g and g'_0 separated, the nonzero terms have g in the $(q+1)$st entry and are independent of $g'_0 \in N$ since $\varphi - \delta\gamma_q \in F^pA^*$. Hence

$$(\varphi - \delta\gamma_q)(g^{(1)}, \ldots, g^{(q)}, gg'_0, g_1, \ldots, g_p)$$

is independent of g'_0, so $\varphi - \delta\gamma_q \in F^{p+1}A^* \cap A^{p+q+1}$. Thus we may choose $v = \gamma_q$ and conclude that the homomorphism $\Phi^{p,q}$ is a monomorphism.

To prove $\Phi^{p,q}$ is an epimorphism, we must show that for any $u \in \operatorname{Hom}_{G/N}(\overline{X}_p, \operatorname{Hom}_N(X_{N,q}, A))$ such that $u[\bar{g}_1, \ldots, \bar{g}_p]$ is a q-cocycle of N in A, there is an element $h \in F^pA^* \cap A^{p+q}$ such that $\delta h \in F^{p+1}A^* \cap A^{p+q+1}$ and $\rho_p h = u$. Thus, for a given u define $\gamma\colon X_{N,q} \times X_p \to A$ by

$$\gamma(g'_1, \ldots, g'_q, g_1, \ldots, g_p) = (u[\bar{g}_1, \ldots, \bar{g}_p])(g'_1, \ldots, g'_q).$$

If $q = 0$, we may choose $h = \gamma$. If $q > 0$, construct the same extensions $\gamma_1, \ldots, \gamma_q$ of γ where the map φ must be replaced by zero here. Thus we obtain an extension γ_q of γ such that

$$\gamma_q \in F^pA^* \cap A^{p+q} \qquad \text{and} \qquad \delta\gamma_q \in F^{p+1}A^* \cap A^{p+q+1}.$$

Choose $h = \gamma_q$, then $\rho_p h = u$ and $\Phi^{p,q}$ is an epimorphism.

53.3 *The term* $E_2^{p,q}$. To compute $E_2^{p,q}$ we must determine the relationship between the differential homomorphism $d_1^{p,q}$ of the Hochschild–Serre spectral sequence and the isomorphism $\Phi^{p,q}$ of Section 53.2. We will show

$$\Phi^{p,q}\, d_1^{p-1,q} = (-1)^q \delta \Phi^{p-1,q}.$$

To establish this identity we must obtain certain identities involving partial coboundary operators.

Let $\varphi \in A^{p+q-1}, p > 0, q > 0$. Let $(g^{(1)}, \ldots, g^{(q)}, g_1, \ldots, g_p)$ denote a $p+q$-tuple of elements of the group G. Define the partial coboundary operators δ_q, ∂_p by

$$
\begin{aligned}
&(\delta_q \varphi)(g^{(1)}, \ldots, g^{(q)}, g_1, \ldots, g_p) \\
&\quad = g^{(1)} \varphi(g^{(2)}, \ldots, g^{(q)}, g_1, \ldots, g_p) \\
&\qquad + \sum_{i=1}^{q-1} (-1)^i \varphi(g^{(1)}, \ldots, g^{(i)} g^{(i+1)}, \ldots, g^{(q)}, g_1, \ldots, g_p) \\
&\qquad + (-1)^q \varphi(g^{(1)}, \ldots, g^{(q-1)}, g_1, \ldots, g_p)
\end{aligned}
$$

and

$$
\begin{aligned}
&(\partial_p \varphi)(g^{(1)}, \ldots, g^{(q)}, g_1, \ldots, g_p) \\
&\quad = g_1 \varphi(g_1^{-1} g^{(1)} g_1, \ldots, g_1^{-1} g^{(q)} g_1, g_2, \ldots, g_p) \\
&\qquad + \sum_{i=1}^{p-1} (-1)^i \varphi(g^{(1)}, \ldots, g^{(q)}, g_1, \ldots, g_i g_{i+1}, \ldots, g_p) \\
&\quad + (-1)^p \varphi(g^{(1)}, \ldots, g^{(q)}, g_1, \ldots, g_{p-1}).
\end{aligned}
$$

Let $S = (s_1, \ldots, s_p)$ be an increasing subsequence of the sequence $(1, 2, \ldots, p+q)$, and let $S^* = (s_1^*, \ldots, s_q^*)$ be the complement of S written in the increasing order. Set $b_0 = 1$, $b_i = g_1 \cdot \cdots \cdot g_i$. For $1 \leqq j \leqq q$ write $j^* = s_j^* - j$ (which is the number of $s_i < s_j^*$) and set $v(S) = \sum_{j=1}^q j^*$. For any $\psi \in A^{p+q}$ define

$$\psi_S(g^{(1)}, \ldots, g^{(q)}, g_1, \ldots, g_p) = \psi(h_1, \ldots, h_{p+q})$$

where $h_{s_i} = g_i$ and $h_{s_j^*} = b_{j^*}^{-1} g^{(j)} b_{j^*}$. Define the shuffle of ψ by

$$\mathrm{Sh}_p\, \psi = \sum_S (-1)^{v(S)}\, \psi_S$$

where S runs over the set of all increasing subsequences of $(1, 2, \ldots, p+q)$ of length p—for example, when $q = 1$ and $p = 2$,

$$
\begin{aligned}
\mathrm{Sh}_2\, \psi(g^{(1)}, g_1, g_2) &= \psi(g^{(1)}, g_1, g_2) - \psi(g_1, g_1^{-1} g^{(1)} g_1, g_2) \\
&\quad + \psi(g_1, g_2, (g_1 g_2)^{-1} g^{(1)} (g_1 g_2)).
\end{aligned}
$$

For $\varphi \in A^{p+q-1}$ we have

$$\mathrm{Sh}_p(\delta\varphi) = \delta_q(\mathrm{Sh}_p\, \varphi) + (-1)^q\, \partial_p(\mathrm{Sh}_{p-1}\, \varphi). \tag{12}$$

To verify this identity evaluate the left-hand side at $(g^{(1)}, \ldots, g^{(q)}, g_1, \ldots, g_p)$ using the definition of Sh_p and the coboundary homomorphism. After some cancellations the left-hand side reduces to the values of φ at $(p + q - 1)$-tuples whose entries are of type

$$b_{j*}^{-1}g^{(j)}\,b_{j*}g_{j*+1},\qquad b_{j*}^{-1}g^{(j)}g^{(j+1)}\,b_{j*},\qquad g_i,\qquad \text{and}\qquad g_ig_{i+1}.$$

Furthermore, these terms of the left-hand side are in one-to-one correspondence with the terms of the right-hand side. We will omit the verification that the corresponding terms carry the same signs (see [51, p. 124]).

EXERCISE Verify (12) for the case $q = 1$ and $p = 2$.

Let $\varphi \in F^{p-1}A^* \cap A^{p+q-1}$ such that $\delta\varphi \in F^pA^* \cap A^{p+q}$. Let ρ_p: $F^pA^* \to \mathrm{Hom}_{G/N}(\bar{X}_p, \mathrm{Hom}_N(X_{N,*}, A))$ be the morphism of complexes defined in 53.1. It follows from the definitions that if $\gamma \in F^pA^* \cap A^{p+q}$, the restriction of the first q entries in $\mathrm{Sh}_p\,\gamma$ to N yields the image of $\rho\gamma$ in $\mathrm{Hom}_G(X_p, \mathrm{Hom}(X_{N,q}, A))$. Hence if the identity for φ above is evaluated at $(g'_1, \ldots, g'_q, g_1, \ldots, g_p)$ where $g'_j \in N$ and $g_i \in G$, we obtain

$$\rho_p(\delta\varphi)(\bar{g}_1, \ldots, \bar{g}_p) = \delta(h(g_1, \ldots, g_p)) + (-1)^q\delta(\rho_{p-1}(\varphi))(\bar{g}_1, \ldots, \bar{g}_p)$$

where $h(g_1, \ldots, g_p)(g'_1, \ldots, g'_{q-1}) = \mathrm{Sh}_p\;\varphi(g'_1, \ldots, g'_{q-1}, g_1, \ldots, g_p)$. Hence if $\bar{\varphi} \in E_1^{p-1,q}$ corresponds to φ, then

$$\Phi^{p,q}(d_1^{p-1,q}(\bar{\varphi})) = (-1)^q\delta(\Phi^{p-1,q}(\bar{\varphi})).$$

This identity implies the following:

The isomorphism $\Phi^{p,q}$: $E_1^{p,q} \cong \mathrm{Hom}_{G/N}(\bar{X}_p, H^q(N, A))$ induces an isomorphism $E_2^{p,q} \cong H^p(G/N, H^q(N, A))$.

54. Exact Sequences Involving the Terms of Spectral Sequences

In this section we return to the general discussion of a graded differential group A^* with a filtration. As always we assume the filtration is such that for each m and large enough n we have $(F^{p-n}A)^m = A^m$ and $(F^{p+n}A)^m = 0$. Hence as in 49.7 the filtration of A^* induces a filtration of $H(A^*)$. Recall that $E_{r+1}^{p,q} = \ker d_r^{p,q}/\mathrm{im}\; d_r^{p-r,q+r-1}$, so that if $E_r^{p,q} = 0$, then $E_s^{p,q} = 0$ for all $s \geqq r$.

54.1 *If $r < s \leqq \infty$ and $E_r^{m,n} = 0$ for $m + n = p + q - 1$, and $p - s < m \leqq p - r$, then there exists a monomorphism $E_s^{p,q} \to E_r^{p,q}$.*

Proof Since $E_r^{p-r,q+r-1} = 0$, its image under $d_r^{p-r,q+r-1}$ is zero. Hence $E_{r+1}^{p,q} = \ker d_r^{p,q}$, so $E_{r+1}^{p,q} \subseteq E_r^{p,q}$. Using induction suppose for $t > 0$ we have $E_{r+t}^{p,q} \subseteq E_r^{p,q}$. Since by the remarks above $E_r^{p-r-t,q+r+t-1} = 0$ implies $E_{r+t}^{p-r-t,q+r+t-1} = 0$, we have $d_{r+t}^{p-r-t,q+r+t-1} = 0$. Hence $E_{r+t+1}^{p,q} = \ker d_{r+t}^{p,q}$, so $E_{r+t+1}^{p,q} \subseteq E_{r+t}^{p,q}$. Therefore $\cdots \subset E_{r+t}^{p,q} \subset \cdots \subset E_r^{p,q}$. Since in this case $E_\infty^{p,q} = \bigcap_{t=0}^{\infty} E_{r+t}^{p,q}$, we also have $E_\infty^{p,q} \subseteq E_r^{p,q}$.

54.2 *If $r < s \leqq \infty$ and $E_r^{m,n} = 0$ for $m + n = p + q + 1$ and $p + r \leqq m < p + s$, then there exists an epimorphism $E_r^{p,q} \to E_s^{p,q}$.*

Proof The proof is dual to the proof of 54.1 when s is finite. For $s = \infty$ the result follows from the fact that $E_\infty^{p,q}$ is the limit of the sequence of epimorphisms $E_r^{p,q} \to E_{r+1}^{p,q} \to \cdots$.

54.3 *If $E_\infty^{m,n-m} = 0$ for $m < p$, then $F^pH^n(A^*) = H^n(A^*)$, and there exists an epimorphism $H^n(A^*) \to E_\infty^{p,n-p}$.*

Proof Since $H^n(A^*) = \bigcup F^mH^n(A^*)$ and the hypothesis implies $F^mH^n(A^*) = F^{m+1}H^n(A^*)$ for $m < p$, we have $H^n(A^*) = F^pH^n(A^*)$. The epimorphism exists because $E_\infty^{p,n-p} \cong F^pH^n(A^*)/F^{p+1}H^n(A^*)$.

54.4 *If $E_\infty^{m,n-m} = 0$ for $m > p$, then $F^{p+1}H^n(A^*) = 0$, and there exists a monomorphism $E_\infty^{p,n-p} \to H^n(A^*)$.*

The proof is similar to the proof of 54.3.

55. A Decomposition of $H^p(G, A)$ by Means of a Normal Hall Subgroup of G

Let $N \trianglelefteq G$. We have seen in Sections 52 and 53 that the Cartan–Leray or the Hochschild–Serre spectral sequences result in

$$H^p(G/N, H^q(N, A)) \underset{p}{\Rightarrow} H^{p+q}(G, A).$$

Recall that in this chapter $H^0(G, M)$ stands for M^G. Then by 54.3 and 54.1

we have the homomorphisms

$$(13) \qquad H^q(G, A) \to E_\infty^{0,q} \to E_2^{0,q} \cong H^q(N, A)^{G/N}$$

where the first is an epimorphism and the second is a monomorphism. Similarly, by 54.4 and 54.2 we have

$$(14) \qquad H^p(G/N, A^N) \cong E_2^{p,0} \to E_\infty^{p,0} \to H^p(G, A)$$

where the first map is an epimorphism and the second a monomorphism.

55.1 *Let G be a finite group and N a normal Hall subgroup of G (N is Hall if its order and index are relatively prime). Let $(G: N) = m$ and $(N: 1) = n$. Then for $p > 0$,*

$$H^p(G, A) \cong H^p(N, A)^{G/N} \oplus H^p(G/N, A^N).$$

PROOF By 15.5, for any G-module A, the order of each element of $H^p(G, A)$ is a divisor of the order of G. Therefore if $\tau \in H^p(G/N, H^q(N, A))$, then $n\tau = 0$ if $q > 0$, and $m\tau = 0$ if $p > 0$. Hence the same holds for any element of $E_2^{p,q}$ and so for any element of $E_r^{p,q}$ if $r \geqq 2$. This implies $E_r^{p,q} = 0$ for $r \geqq 2$, $p > 0$, $q > 0$ since any element $\tau \in E_r^{p,q}$ can be written $\tau = 1\tau = (am + bn)\tau$ for some integers a and b. Therefore the map $d_r^{p,q}: E_r^{p,q} \to E_r^{p+r,q-r+1}$ is zero unless $q = r - 1$ and $p = 0$. But for any $\tau \in E_r^{0,r-1}$ we have $nd_r^{0,r-1}\tau = d_r^{0,r-1}(n\tau) = 0$, and $md_r^{0,r-1}\tau = 0$ since $d_r^{0,r-1}\tau \in E^{r,0}$. Hence

$$d_r^{0,r-1}\tau = d_r^{0,r-1}(am + bn)\tau = 0,$$

and so $d_r^{p,q} = 0$ for $r \geqq 2$ and all p, q. This shows that the maps $E_2^{p,0} \to E_\infty^{p,0}$ and $E_\infty^{0,q} \to E_2^{0,q}$ in (13) and (14) above are isomorphisms. Let r_q denote the composite map in (13) and l_p the composite map in (14). Then under the hypotheses here r_q is an epimorphism and l_p is a monomorphism. Furthermore, $F^1H^p(G, A)$, the kernel of r_p, is equal to $F^p(G, A)$, the image of l_p. Therefore the sequence

$$(15) \qquad 0 \to H^p(G/N, A^N) \xrightarrow{l_p} H^p(G, A) \xrightarrow{r_p} H^p(N, A)^{G/N} \to 0$$

is exact. It is enough to show this sequence splits. Let $am = e_1$ and $bn = e_2$ where $am + bn = 1$. Then for any $\tau \in H^p(G, A)$, $e_1e_2\tau = e_2e_1\tau = abmn\tau = 0$ since mn is the order of G, and

$$e_1\tau = am\cdot 1\tau = (a^2m^2 + abmn)\tau = a^2m^2\tau = e_1^2\tau.$$

Similarly $e_2\tau = e_2^2\tau$. Hence e_1 and e_2 are orthogonal generating idempotents for the **Z**-module $H^p(G, A)$ (see [20]). Therefore

$$H^p(G, A) = e_1 H^p(G, A) \oplus e_2 H^p(G, A).$$

Moreover $r_p e_2 H^p(G, A) = 0$, $r_p e_1 H^p(G, A) = H^p(N, A)^{G/N}$, and $l_p H^p(G/N, A^N) = e_2 H^p(G, A)$. This implies that the sequence (15) splits and completes the proof of 55.1.

56. The Five-Term Exact Sequence of Hochschild and Serre

Let A be a G-module and $N \trianglelefteq G$. If $H^i(N, A) = 0$ for $0 < i < n$, then the sequence

$$0 \to H^n(G/N, A^N) \xrightarrow{l_n} H^n(G, A) \xrightarrow{r_n} H^n(N, A)^{G/N} \xrightarrow{t_{n+1}} H^{n+1}(G/N, A^N) \xrightarrow{l_{n+1}} H^{n+1}(G, A)$$

is exact where l_n, r_n are the homomorphisms in Section 55, and t_{n+1} corresponds canonically to $d_{n+1}: E_{n+1}^{0,n} \to E_{n+1}^{n+1,0}$.

PROOF Since $E_2^{1,0} \cong E_\infty^{1,0} \cong F^1 H^1(G, A)$, the map l_1, which is the composite map

$$H^1(G/N, A^N) \cong E_2^{1,0} \to E_\infty^{1,0} = H^1(G, A),$$

is a monomorphism. By induction suppose l_i is a monomorphism for $1 \leq i < m$. There is a row-exact commutative diagram

$$\begin{array}{ccccccccc} 0 & \to & A & \to & B & \to & C & \to & 0 \\ & & \uparrow & & \uparrow & & \uparrow & & \\ 0 & \to & A^N & \to & B^N & \to & C^N & \to & 0 \end{array}$$

where B (resp. B^N) is G-induced (resp. G/N-induced) (17.1, 35.4, 35.5, and the proof of 35.6). As in the proof of 35.6 the diagram

$$\begin{array}{ccc} H^{n-1}(G, C) & \overset{\delta}{\cong} & H^n(G, A) \\ \uparrow l_{n-1} & & \uparrow l_n \\ H^{n-1}(G/N, C^N) & \overset{\delta}{\cong} & H^n(G/N, A^N) \end{array}$$

is commutative. The induction hypothesis applied to the G-module C implies l_{n-1} in (16) is a monomorphism. Hence l_n: $H^n(G/N, A^n) \to H^n(G, A)$ is a monomorphism. To show exactness at $H^n(G, A)$ we note that

$$\operatorname{im} l_n = F^m H^m(G, A) \qquad \text{and} \qquad \ker r_n = F^1 H^m(G, A).$$

Moreover our hypothesis implies $E_r^{p,q} = 0$ for $0 < q < n$ and $r \geqq 2$. Hence for $p = n - q$ and $r = n + 1$ we have

$$0 = E_{n+1}^{n-q,q} = E_\infty^{n-q,q} \cong F^{n-q}H^n(G, A)/F^{n-q+1}H^n(G, A)$$

for any $0 < q < n$. Hence

$$F^1 H^n(G, A) = F^2 H^n(G, A) = \cdots = F^n H^n(G, A).$$

To show exactness at $H^n(N, A)^{G/N}$ we first define the map t_{n+1}, precisely, as the composite map

$$H^n(N, A)^{G/N} \simeq E_2^{0,n} \simeq E_{n+1}^{0,n} \xrightarrow{d_{n+1}} E_{n+1}^{n+1,0} \simeq E_2^{n+1,0} \simeq H^{n+1}(G/N, A^N)$$

where the canonical isomorphism $E_2^{0,n} \simeq E_{n+1}^{0,n}$ results from the fact that $d_r(E_r^{0,n}) \subset E_r^{r,n+1-r} = 0$ if $2 \leqq r \leqq n$, and the canonical isomorphism $E_{n+1}^{n+1,0} \simeq E_2^{n+1,0}$ results from $E_r^{n+1-r,r-1} = 0$ for $2 \leqq r \leqq n$. The remaining isomorphisms exist in general for the Hochschild–Serre spectral sequence. Hence $\ker t_{n+1}$ is the image of $E_{n+2}^{0,n}$ in $H^n(N, A)^{G/N}$. Since $E_{n+2}^{0,n}$ is isomorphic to $H^n(G, A)/F^1H^n(G, A) \cong E^{0,n} = \operatorname{im} r_n$, the sequence is exact at $H^n(N, A)^{G/N}$. To show exactness at $H^{n+1}(G/N, A^N)$ we observe that $\operatorname{im} t_{n+1} \cong d_{n+1}(E_{n+1}^{0,n})$ which is the kernel of $E_{n+1}^{n+1,0} \to E_{n+2}^{n+1,0} \cong F^{n+1}H^{n+1}(G, A)$. Therefore the image of t_{n+1} is the kernel of l_{n+1}. This completes the proof.

Remarks

(i) In case $n = 1$, the hypothesis in Section 56 is vacuous. Hence we also have the exact sequence

$$(17) \qquad 0 \to H^1(G/N, A^N) \to H^1(G, A) \to H^1(N, A)^{G/N} \to H^2(G/N, A^N) \to H^2(G, A).$$

(ii) Let $H^0(S, A)$ be the monoid cohomology for any $S \leqq G$, i.e., $H^0(S, A)$ is the group of 0-cocycles of S in A. Let $T_S = \sum_{g \in S} g$ be the trace as in Chapter 2. Then for $N \trianglelefteq G$,

$$0 \to T_N A \to A^N \to H^0(N, A) \to 0$$

may be viewed as an exact sequence of G/N-homomorphisms. Hence we have the exact sequence

$$0 \to (T_N A)^{G/N} \to (A^N)^{G/N} \to H^0(N, A)^{G/N} \to H^1(G/N, T_N A).$$

Since $(A^N)^{G/N} = A^G$ we have the exact sequence

$$0 \to H^0(G/N, T_N A) \to H^0(G, A) \to H^0(N, A)^{G/N} \to H^1(G/N, T_N A)$$

which is analogous to the Hochschild–Serre sequence.

(iii) Let $N \trianglelefteq G$, A a G-module, and $H^i(N, A) = 0$ for $0 < i < n$. Then the five-term Hochschild–Serre exact sequence may be compared with five terms of the long exact sequence obtained from the exact sequence of G-homomorphisms

$$0 \to A \to \mathrm{Hom}_N(\mathbf{Z}G, A) \to \mathrm{Hom}_N(\mathbf{Z}G, A)/A \to 0$$

where $\mathrm{Hom}_N(\mathbf{Z}G, A)$ is given the G-module structure by $g\varphi(g_1) = \varphi(g_1 g)$, for $\varphi \in \mathrm{Hom}_N(\mathbf{Z}G, A)$, and A is identified with the submodule $\mathrm{Hom}_G(\mathbf{Z}G, A)$. Berkson and McConnell [10] have shown that if $H^i(N, A) = 0$ for $0 < i < n$, there exist homomorphisms α, β, γ such that the row-exact diagram

$$\begin{array}{ccccccccccc}
0 \to & H^n(G/N, A^N) & \to & H^n(G, A) & \to & H^n(N, A)^{G/N} & \xrightarrow{-t_{n+1}} & H^{n+1}(G/N, A^N) & \to & H^{n+1}(G, A) \\
& \downarrow \alpha & & \downarrow \text{id.} & & \downarrow \beta & & \downarrow \gamma & & \downarrow \text{id.} \\
0 \to & H^{n-1}(G, \mathrm{Hom}_N(\mathbf{Z}G, A)/A) & \to & H^n(G, A) & \to & H^n(G, \mathrm{Hom}_N(\mathbf{Z}G, A)) & \to & H^n(G, \mathrm{Hom}_N(\mathbf{Z}G, A)/A) & \to & H^{n+1}(G, A)
\end{array}$$

is commutative, α is an isomorphism, and β and γ are monomorphisms.

(iv) An element $x \in H^{i-1}(N, A)$ is called *transgressive* if there is a cochain $\varphi \in \mathrm{Hom}_G(X_{i-1}, A)$ whose restriction to $X_{N,i-1}$ is a cocycle representing x and which is such that $d\varphi$ is the image in $\mathrm{Hom}_G(X_i, A)$ of a cocycle in $\mathrm{Hom}_{G/N}(\overline{X}_i, A^N)$. The transgressive elements of $H^{i-1}(N, A)$ are the canonical image of $E_i^{0,i-1}$ in $H^{i-1}(N, A)$ [51, Section 3, Chapter III]. In the five-term exact sequence above, the transgressive elements of $H^n(N, A)$ are the elements in $H^n(N, A)^{G/N}$, and the map t_{n+1} there is called the *transgression*.

57. Homology Spectral Sequences

The spectral sequences developed thus far in this chapter are related to cohomology groups. In this section we outline some of the dual theory needed in Chapter 9 for homology groups.

57.1 The exact couples $\langle D, E, i, j, k\rangle$ in this section are bigraded and the homomorphisms i, j, k are homogeneous as before. But in this case i is of bidegree $(1, -1)$, j is of bidegree $(0, 0)$, and k is of bidegree $(-1, 0)$. Hence dualizing the results of Section 49,

i^n is of bidegree $(1, -1)$,
j^n is of bidegree $(-n + 1, n - 1)$,
k^n is of bidegree $(-1, 0)$,
d^n is of bidegree $(-n, n - 1)$.

The exact couple $\langle D, E, i, j, k\rangle$ is called *regular* if $D_{p,q} = 0$ for $p < 0$, and $E_{p,q} = 0$ for $q < 0$. Moreover the exactness of $D_{p,q} \xrightarrow{j} E_{p,q} \xrightarrow{k} D_{p-1,q}$ implies $E_{p,q} = 0$ for $p < 0$. Therefore for $n > 0$ we have $E^n_{p,q} = 0$ if $p < 0$, $q < 0$.

57.2 As in 49.6, there is an exact couple associated with a filtration of a chain complex A_*. The groups $E^n_{p,q}$ and $E^\infty_{p,q}$ can be defined as in 49.6 and 49.7. If the filtration F of the complex A_* is such that for large enough n, $(F^{-n}A)_* = 0$ and $(F^nA)_* = A_*$, then $E^\infty_{p,q} \cong E^0_{p,q}H(A_*)$. If the filtration of the chain complex is such that the resulting exact couple is regular, then as in (6) of 50.2 we have the exact sequence

$$H_2(A_*) \to E^2_{2,0} \to E^2_{0,1} \to H_1(A_*) \to E^2_{1,0} \to 0. \tag{18}$$

57.3 If N is a normal subgroup of G, and M is a G-module, there is a spectral sequence dual to the Cartan–Leray spectral sequence which results in

$$H_p(G/N, H_q(N, M)) \underset{p}{\Rightarrow} H_{p+q}(G, M).$$

Therefore by (18) we have the exact sequence

$$\begin{aligned} H_2(G, A) &\to H_2(G/N, H_0(N, M)) \to H_0(G/N, H_1(N, M)) \\ &\to H_1(G, M) \to H_1(G/N, H_0(N, M)) \to 0. \end{aligned} \tag{19}$$

CHAPTER

9

Descending Central Series

This chapter contains group theoretic results that are proven with the aid of the five-term Hochschild–Serre exact sequence and descending central series of groups.

58. A Homological Condition for Isomorphism of Nilpotent Groups

In this section G is a group (not necessarily finite). For $p = 0$, $\mathbf{Z}_p$ is the trivial G-module $\mathbf{Z}$, and for p a prime $\mathbf{Z}_p$ is the trivial G-module $\mathbf{Z}/p\mathbf{Z}$.

For each group G we define the descending sequence of normal subgroups G_n as follows:

$$G_0 = G, \qquad G_{n+1} = (G_n)^p[G, G_n] \qquad \text{for} \quad n \geqq 0,$$

$$G_\infty = \bigcap_{n=0}^{\infty} G_n,$$

where $(G_n)^p[G, G_n]$ is the subgroup of G generated by all elements of the form $g_n^p \cdot ghg^{-1}h^{-1}$, g_n, $h \in G_n$, $g \in G$. The sequence

$$G = G_0 \geqq G_1 \geqq \cdots$$

is a descending central series, and for $p = 0$ it is the lower central series of G.

Let N be a normal subgroup of G. Then we have the spectral sequence of Section 57 which results in

$$H_m(G/N, H_n(N, \mathbf{Z}_p)) \underset{m}{\Rightarrow} H_{m+n}(G, \mathbf{Z}_p).$$

Since $\mathbf{Z}_p$ is G-trivial, the exact sequence (19) of 57.3 gives the exact sequence

$$\begin{aligned} H_2(G, \mathbf{Z}_p) \to H_2(G/N, \mathbf{Z}_p) \to N/N^p[G, N] \to H_1(G, \mathbf{Z}_p) \\ \to H_1(G/N, \mathbf{Z}_p) \to 0 \end{aligned} \tag{1}$$

where the third term is obtained by

$$H_0(G/N, H_1(N, \mathbf{Z}_p)) \cong H_0(G/N, N/N^p[N, N]) = N/N^p[G,N].$$

58.1 *Let $\varphi: G \to G'$ be a homomorphism of groups which induces an isomorphism $H_1(G, \mathbf{Z}_p) \cong H_1(G', \mathbf{Z}_p)$ and an epimorphism $H_2(G, \mathbf{Z}_p) \to H_2(G', \mathbf{Z}_p)$. If for some n, φ induces an isomorphism $G/G_n \cong G'/G'_n$, then φ induces the isomorphisms $G_n/G_{n+1} \cong G'_n/G'_{n+1}$ and $G/G_{n+1} \cong G'/G'_{n+1}$.*

PROOF Using the exact sequence (1) in the case where $N = G_n$, and again with G, N replaced by G' and the normal subgroup G'_n we obtain the commutative diagram

$$\begin{array}{ccccccccccc} H_2(G, \mathbf{Z}_p) & \to & H_2(G/G_n, \mathbf{Z}_p) & \to & G_n/G_{n+1} & \to & H_1(G, \mathbf{Z}_p) & \to & H_1(G/G_n, \mathbf{Z}_p) & \to & 0 \\ \downarrow & & \downarrow & & \downarrow & & \downarrow & & \downarrow & & \\ H_2(G', \mathbf{Z}_p) & \to & H_2(G'/G'_n, \mathbf{Z}_p) & \to & G'_n/G'_{n+1} & \to & H_1(G', \mathbf{Z}_p) & \to & H_1(G'/G'_n, \mathbf{Z}_p) & \to & 0 \end{array}$$

where the vertical maps are induced by φ. The first vertical map is an epimorphism, the second, fourth, and fifth are isomorphisms by the hypothesis. Therefore by the Five Lemma the third vertical map is an isomorphism. To show we also have $G/G_{n+1} \cong G'/G'_{n+1}$ apply the Five Lemma to the commutative diagram

$$\begin{array}{ccccccccc}
1 & \to & G_n/G_{n+1} & \to & G/G_{n+1} & \to & G/G_n & \to & 1 \\
 & & \downarrow\wr & & \downarrow & & \downarrow\wr & & \\
1 & \to & G'_n/G'_{n+1} & \to & G'/G'_{n+1} & \to & G'/G'_n & \to & 1
\end{array}$$

where the second vertical map is induced by φ.

58.2 *Let $\varphi\colon G \to G'$ be a homomorphism of groups. If φ induces the isomorphisms $G/G_n \cong G'/G'_n$ for every n, then φ induces a monomorphism $\varphi_\infty\colon G/G_\infty \to G'/G'_\infty$.*

PROOF By the universal mapping property of inverse limits, the maps $\alpha_n\colon G/G_\infty \to G/G_n$ induce a map $\alpha\colon G/G_\infty \to \varprojlim_n G/G_n$ such that the diagram

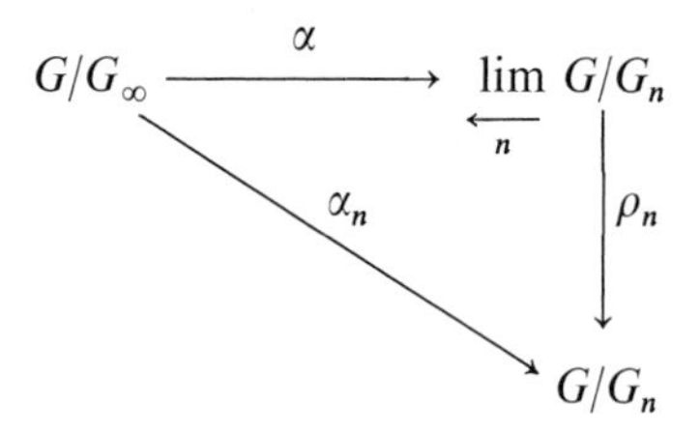

is commutative for every n; α is a monomorphism, for, suppose $\alpha g G_\infty = 0$. Then $g \in G_n$ for every n, so $g \in G_\infty$. Consider the diagram

$$\begin{array}{ccc}
G/G_\infty & \xrightarrow{\ \varphi_\infty\ } & G'/G'_\infty \\
\Big\downarrow \alpha & & \Big\downarrow \alpha' \\
\varprojlim_n G/G_n & \xrightarrow{\ \beta\ } & \varprojlim_n G'/G'_n
\end{array}$$

where φ_∞ is induced by φ, and β is induced by the family of maps

$$\varprojlim_n G/G_n \xrightarrow{\rho_n} G/G_n \xrightarrow{\sim} G'/G'_n.$$

The composite map $\beta\alpha$ in the diagram is a monomorphism. Hence φ_∞ is a monomorphism since the diagram is commutative.

The following example by Stallings [100] shows that φ_∞ need not be an epimorphism. Let $G = G'$ be free on the letters x and y, and let $\varphi: G \to G$ be defined by $\varphi(x) = x$, $\varphi(y) = y[x, y]$. Let $p = 0$; then φ induces an automorphism of $H_1(G, \mathbf{Z})$ and $H_2(G, \mathbf{Z}) = 0$. Beginning with $n = 0$ and proceeding with induction it follows that φ induces an isomorphism $\varphi_n: G/G_n \cong G/G_n$. Since in a free group $\bigcap_{n=0}^{\infty} G_n = 1$ (see 60.3), $\varphi_\infty = \varphi$ and φ is not an epimorphism since y is not in the image of φ. We will now give a condition under which φ_∞ is an epimorphism.

58.3 *Let $\varphi: G \to G'$ be an epimorphism. If $\varphi_n: G/G_n \cong G'/G'_n$ is an isomorphism, then $\varphi_\infty: G/G_\infty \to G'/G'_\infty$ is an isomorphism.*

PROOF By 58.2, φ_∞ is a monomorphism. To show it is an epimorphism we use the fact that $G \overset{\varphi}{\to} G' \to G'/G'_\infty$ is an epimorphism and this composite map is equal to the composite map $G \to G/G_\infty \to G'/G'_\infty$.

58.4 *Let $\varphi: G \to G'$ be a homomorphism which induces $H_1(G, \mathbf{Z}_p) \cong H_1(G', \mathbf{Z}_p)$ and $H_2(G, \mathbf{Z}_p) \to H_2(G', \mathbf{Z}_p)$ is an epimorphism. Then φ induces an isomorphism $G/G_n \cong G'/G'_n$ and a monomorphism $G/G_\infty \to G'/G'_\infty$. If in addition φ is an epimorphism, then it induces an isomorphism $G/G_\infty \cong G'/G'_\infty$.*

PROOF Follows by induction on n and 58.1–58.3.

58.5 (Stallings [100]) *Let $\varphi: G \to G'$ be a homomorphism of nilpotent groups which induces $H_1(G, \mathbf{Z}) \cong H_1(G', \mathbf{Z})$ and $H_2(G, \mathbf{Z}) \to H_2(G', \mathbf{Z})$ is an epimorphism. Then φ is an isomorphism.*

PROOF Follows from 58.4 and the fact that the lower central series of a nilpotent group is finite, i.e., for some n, $G_n = \{1\}$.

59. Existence of a Normal p-Complement

In this section G is a finite group and $G = G_0 > G_1 > \cdots$ is the descending central series defined in Section 58.

59.1 (Tate [107]) *Let G be a finite group, $S \leqq G$ such that $(G: S)$ is prime to p. If $S_1 = S \cap G_1$, then $S_n = S \cap G_n$ for all n, and $S_\infty = S \cap G_\infty$.*

PROOF The proof is by induction on n. Suppose we have already shown $S_n = S \cap G_n$. Then by the five-term Hochschild–Serre exact sequence we have the commutative diagram

$$\begin{array}{ccccccccc}
0 \to H^1(G/G_n, \mathbf{Z}_p) & \to & H^1(G, \mathbf{Z}_p) & \to & H^1(G_n, \mathbf{Z}_p)^{G/G_n} & \xrightarrow{t} & H^2(G/G_n, \mathbf{Z}_p) & \to & H^2(G, \mathbf{Z}_p) \\
\downarrow \alpha & & \downarrow \beta & & \downarrow \gamma & & \downarrow \delta & & \downarrow \varepsilon \\
0 \to H^1(S/S_n, \mathbf{Z}_p) & \to & H^1(S, \mathbf{Z}_p) & \to & H^1(S_n, \mathbf{Z}_p)^{S/S_n} & \xrightarrow{t} & H^2(S/S_n, \mathbf{Z}_p) & \to & H^2(S, \mathbf{Z}_p)
\end{array}$$

where the vertical maps are restrictions; α and δ are isomorphisms by the induction hypothesis and the fact that $S/S \cap G_n \cong SG_n/G_n = G/G_n$. Since $\mathbf{Z}_p$ is uniquely divisible by the index of S in G and, by 32.6, $\mathrm{cor}_{S,G}\ \mathrm{res}_{G,S} = (G: S)$, β and ε are monomorphisms because they are restriction maps. We contend the map β is an isomorphism. Namely, since $S/S_1 \cong G/G_1$ we have

$$\begin{aligned} H^1(G, \mathbf{Z}_p) &\cong \mathrm{Hom}(G, \mathbf{Z}_p) \cong \mathrm{Hom}(G/G_1, \mathbf{Z}_p) \cong \mathrm{Hom}(S/S_1, \mathbf{Z}_p) \\ &\cong H^1(S, \mathbf{Z}_p). \end{aligned}$$

From the Five Lemma it follows that γ is an isomorphism. This implies

$$\begin{aligned} \mathrm{Hom}(G_n/G_{n+1}, \mathbf{Z}_p) &= \mathrm{Hom}(G_n/G_n^p[G, G_n], \mathbf{Z}_p) \\ &\cong \mathrm{Hom}_{G/G_n}(G_n/[G_n, G_n], \mathbf{Z}_p) \cong H^1(G_n, \mathbf{Z}_p)^{G/G_n} \\ &\cong H^1(S_n, \mathbf{Z}_p)^{S/S_n} \cong \mathrm{Hom}(S_n/S_{n+1}, \mathbf{Z}_p). \end{aligned}$$

Hence by duality $G_n/G_{n+1} = S_n/S_{n+1}$. This implies $S_{n+1} = S_n \cap G_{n+1} = S \cap G_{n+1}$. Thus by induction we have shown $S_n = S \cap G_n$ for every n. Moreover

$$S_\infty = \bigcap_{n=0}^{\infty} (S \cap G_n) = S \bigcap_{n=0}^{\infty} G_n = S \cap G_\infty.$$

The following is a special case of 59.1:

59.2 (Tate [107]) *Let p be a prime and S a Sylow p-subgroup of G. If*

$S_1 = S \cap G_1$, then $SG_\infty = G$ and $S \cap G_\infty = 1$, i.e., G_∞ is a normal p-complement in G.

60. A Condition for the Existence of Free Subgroups

Let F be a free group freely generated by the symbols $\{a\}_{\iota \in I}$. Let Φ be the ring of noncommuting power series in the letters $\{\xi_\iota\}_{\iota \in I}$ with coefficients in $\mathbf{Z}_p$. Then for any ξ_ι, $1 + \xi_\iota \in \Phi$ is a unit since its inverse is $\sum_{j=0}^{\infty} (-1)^j \xi_\iota^j$. Let $\alpha: F \to \Phi$ be defined by $\alpha(a_\iota) = 1 + \xi_\iota$. Then α is a homomorphism into the group of units of Φ. For $X \in \Phi$ let $v(X)$ be the minimum degree of the monomials effectively involved in X and set $v(0) = \infty$. Thus $v(X) = 0$ if X is a unit, $v(\xi_\iota) = 1$, and $v(X_1 X_2) = v(X_1) + v(X_2)$.

60.1 *The homomorphism α is an injection.*

PROOF It is enough to show that if W is a nonempty reduced word in F, then $v(\alpha(W) - 1) < \infty$. Let $W = a_{\iota_1}^{n_1} \cdot \cdots \cdot a_{\iota_k}^{n_k}$. Since, for $n > 0$, $\alpha(a_\iota^n) = (1 + \xi_\iota)^n = 1 + \cdots + \xi_\iota^n$,

$$v(\alpha(a^n) - 1) = r \leqq n.$$

Let $\Phi_r = \{X \in \Phi \text{ s.t. } v(X) > r\}$, then Φ_r is a two-sided ideal. So $\alpha(a_\iota^n) \equiv 1 + z\xi_\iota^r \pmod{\Phi_r}$ where $z \neq 0$, and also $\alpha(a_\iota^{-n}) \equiv 1 - z\xi_\iota^r \pmod{\Phi_r}$. Let

$$r_i = v(\alpha(a_{\iota_j}^{n_j}) - 1),$$

then $\alpha(a_{\iota_j}^{n_j}) \equiv 1 \pm z_j \xi_{\iota_j}^{r_j} \pmod{\Phi_{r_j}}$. Thus since α is a homomorphism, the nonzero term $z_1 \cdot \cdots \cdot z_k \xi_{\iota_1}^{r_1} \cdot \cdots \cdot \xi_{\iota_k}^{r_k}$ will be present in the expression $\alpha(W)$. Hence

$$v(\alpha(W) - 1) \leq \sum_{j=1}^{k} r_j \leq \sum_{j=1}^{k} n_j < \infty.$$

60.2 *Let $\{F_n\}$ be the descending central series for F defined in Section 58. Let W be a nonempty word in F_n. Then $v(\alpha(W) - 1) \geqq n + 1$.*

PROOF We use induction on n. For $n = 0$, $v(\alpha(W) - 1) \geqq 1$ by the definition of α. Suppose the inequality holds for $W \in F_n$. We contend it is enough to prove the inequality for $W = (a')^p a a_n a^{-1} a_n^{-1}$, where a', $a_n \in F_n$ and $a \in F$. For, let $U, V \in F$ such that $v(\alpha(U) - 1) \geqq r$ and $v(\alpha(V) - 1) \geqq r$, i.e., $\alpha(U) \equiv 1 \pmod{\Phi_{r-1}}$ and $\alpha(V) \equiv 1 \pmod{\Phi_{r-1}}$, then

$$\alpha(UV) = \alpha(U)\alpha(V) \equiv 1 \pmod{\Phi_{r-1}}$$

and

$$\alpha(U^{-1}) = (\alpha(U))^{-1} \equiv 1 \pmod{\Phi_{r-1}}$$

so that $v(\alpha(UV) - 1) \geqq r$ and $v(\alpha(U^{-1}) - 1) \geqq r$. Furthermore, using this argument again it suffices to prove the inequality when W is replaced by $(a')^p$ and again by $a a_n a^{-1} a_n^{-1}$. Since $\alpha(a') \equiv 1 + Q \pmod{\Phi_r}$ where Q is a homogeneous polynomial of degree r, we have

$$\alpha((a')^p) = (\alpha(a'))^p \equiv 1 + pQ \pmod{\Phi_r}.$$

So $\alpha((a')^p) \equiv 1 \pmod{\Phi_r}$ since p annihilates the coefficients of Q. This shows $v(\alpha((a')^p) - 1) \geqq r + 1$. To show $v(\alpha(a a_n a^{-1} a_n^{-1}) - 1) \geqq r + 1$ we first observe that $\alpha(a_n) \equiv 1 + Q \pmod{\Phi_r}$ and $\alpha(a_n^{-1}) \equiv 1 - Q \pmod{\Phi_r}$ where Q is a homogeneous polynomial of degree r, and

$$\alpha(a) \equiv 1 + S \pmod{\Phi_r}, \qquad \alpha(a^{-1}) \equiv 1 + T \pmod{\Phi_r}$$

where $v(S) \geqq 1$ and $v(T) \geqq 1$. Then modulo Φ_r

$$\begin{aligned}
\alpha(a a_n a^{-1} a_n^{-1}) &\equiv (1 + S)(1 + Q)(1 + T)(1 - Q) \\
&\equiv 1 + S + Q + T - Q + SQ + ST - SQ + QT - Q^2 \\
&\qquad - TQ + SQT - SQ^2 - STQ - QTQ - SQTQ \\
&\equiv 1 + S + T + ST \\
&\equiv (1 + S)(1 + T) \\
&\equiv \alpha(aa^{-1}) \\
&\equiv 1.
\end{aligned}$$

The results in 60.1 and 60.2 imply the following generalization by Stallings [100] of a theorem of Magnus that the intersection of the lower central series of a free group is trivial.

60.3 *Let F be a free group. Let $G_0 = G$, $G_{n+1} = (G_n)^p[G, G_n]$ for*

$n \geqq 0$ be the subgroup generated by all elements of the form $g_n^p g h g^{-1} h^{-1}$, g_n, $h \in G_n$. Then the intersection of the descending central series $G > G_1 > \cdots$ is trivial.

PROOF If $W \in \bigcap_{n=0}^{\infty} G_n$, then $v(\alpha(W) - 1) > m$ for any integer m by 60.2. Hence, by 60.1, W is the empty word.

For any group G, the homology group $H_1(G, \mathbf{Z}_p)$ is isomorphic to $(G/[G, G]) \otimes \mathbf{Z}_p \cong G/G^p[G, G]$. Let $\theta \colon G \to G/G^p[G, G]$ be the canonical map. We have the following result on free subgroups of G.

60.4 (Stallings [100]) *Let G be a group and p a prime number such that $H_2(G, \mathbf{Z}_p) = 0$. Let $\{g_\iota\}$ be a set of elements of G such that the set $\{\theta g_\iota\}$ is linearly independent over $\mathbf{Z}_p$ where $\theta \colon G \to G/G^p[G, G]$ is the canonical map. Then $\{g_\iota\}$ freely generates a free subgroup of G.*

PROOF Since θ is onto, we may enlarge the set $\{g_\iota\}$ such that its image under θ is a basis of $H_1(G, \mathbf{Z}_p)$ as a vector space over $\mathbf{Z}_p$. We will show this larger set is a basis of a free subgroup of G. Let $\{g_\iota\}$ denote the larger set and let F be the free group freely generated by the letters g_ι. Let $\varphi \colon F \to G$ be the homomorphism $\varphi(g_\iota) = g_\iota$. Then φ induces an isomorphism $H_1(F, \mathbf{Z}_p) \cong H_1(G, \mathbf{Z}_p)$. Since $H_2(G, \mathbf{Z}_p) = 0$ the map $H_2(F, \mathbf{Z}_p) \to H_2(G, \mathbf{Z}_p)$ is onto. So, by 58.4, φ induces a monomorphism $F/F_\infty \to G/G_\infty$. But, by 60.3, $F_\infty = \bigcap_{n=0}^{\infty} F_n$ is trivial, hence $F = F/F_\infty \to G \to G/G_\infty$ is an injection. Therefore $F \to G$ is an injection.

We remark that the result in 60.4 remains true if the coefficient group $\mathbf{Z}_p$ is replaced by the additive group of rational numbers [100, 7.4].

61. Presentations and Free Subgroups

In this section we consider groups presented by n generators and r relations and conditions for the existence of free subgroups with $n - r$ generators, and a result by Magnus [73].

61.1 *Let F be a free group freely generated by the set S. Let I be the* *augmentation ideal of the augmentation map* $\varepsilon: \mathbf{Z}F \to \mathbf{Z}$. *Then* I *is a free* F*-module with the basis* $\{s - 1 : s \in S\}$.

PROOF We have shown in 17.1 that the set $\{f - 1 : f \in F\}$ is a free $\mathbf{Z}$-basis of I. So it is enough to show each $f - 1$ can be written uniquely as a linear combination of the elements $s - 1 \in I$, $s \in S$, with coefficients in $\mathbf{Z}F$. Let $f = x_0 \cdot \cdots \cdot x_n$ where $x_\iota = s$ or $x_\iota = s^{-1}$ for some $s \in S$. For $n = 0$, $f - 1 = x_0 - 1$ so $f - 1 = s - 1$ or $f - 1 = (-s^{-1})(s - 1)$ for some $s \in S$. We contend, for $n \geqq 1$ we have

$$f - 1 = x_0 - 1 + \sum_{\iota=1}^{n} x_0 \cdot \cdots \cdot x_{\iota-1}(x_\iota - 1). \tag{2}$$

This is established by induction on n and the identity

$$f - 1 = x_0 \cdot \cdots \cdot x_{n-1}(x_n - 1) + (x_0 \cdot \cdots \cdot x_{n-1} - 1).$$

To show the representation (2) is unique again we use induction on n. The case $n = 0$ is trivial. For $n \geqq 1$ suppose the representation

$$x_0 \cdot \cdots \cdot x_{n-1} - 1 = x_0 - 1 + \sum_{\iota=1}^{n-1} x_0 \cdot \cdots \cdot x_{\iota-1}(x_\iota - 1)$$

is unique. Let

$$f - 1 = \sum_{\iota=0}^{n} b_\iota(x_\iota - 1), \text{ and also}$$

$$= (x_0 - 1) + \sum_{\iota=1}^{n-1} x_0 \cdot \cdots \cdot x_{\iota-1}(x_\iota - 1) + x_0 \cdot \cdots \cdot x_{n-1}(x_n - 1).$$

This implies

$$\sum_{\iota=0}^{n-1} b_\iota(x_\iota - 1) + (b_n - x_0 \cdot \cdots \cdot x_{n-1})(x_n - 1) = (x_0 - 1) + \sum_{\iota=1}^{n-1} x_0 \cdot \cdots \cdot x_{\iota-1}(x_\iota - 1). \tag{3}$$

Since the right side of (3) is unique by the induction hypothesis, we have $b_0 = 1$, $b_\iota = x_0 \cdot \cdots \cdot x_{\iota-1}$ for $\iota = 1, \ldots, n - 1$, and $b_n - x_0 \cdot \cdots \cdot x_{n-1} = 0$. Hence I is freely generated over F by the elements $s - 1$, $s \in S$.

61.2 *Let F be a free group and M an F-module. Then for $n \geqq 2$ we have* $H^n(G, M) = H_n(G, M) = 0.$

PROOF By 61.1,

$$\cdots \to 0 \to 0 \to I \to \mathbf{Z}F \to \mathbf{Z} \to 0$$

is an F-free resolution of $\mathbf{Z}$. Therefore by the definitions of homology and cohomology of groups (or monoids, see Section 22) we have $H^n(F, M) = 0$ and $H_n(F, M) = 0$ for all $n \geqq 2$.

61.3 (Stammbach [103]) *If G is a group presented by $n + r$ generators and r relations, such that $H_1(G, \mathbf{Z}_p)$ is generated by n elements over $\mathbf{Z}_p$, then:*

(i) *$H_1(G, \mathbf{Z}_p)$ is a vector space of dimension n over $\mathbf{Z}_p$.*
(ii) *$H_2(G, \mathbf{Z}_p) = 0$.*
(iii) *If $g_1, \ldots, g_n$ are elements of G whose images under the canonical map $\theta: G \to G/G_n$ form a basis of G/G_n over $\mathbf{Z}_p$, then $g_1, \ldots, g_n$ freely generate a free subgroup H of G.*
(iv) *For any $n \geqq 0$, the inclusion $H \to G$ induces an isomorphism $H/H_n \cong G/G_n$.*

PROOF Let $1 \to R \to F \to G \to 1$ be a presentation of G where F is freely generated by $n + r$ elements and R by r elements. Then by the exact sequence (1) of Section 58 and 61.2 we have the exact sequence

$$(4) \qquad 0 \to H_2(G, \mathbf{Z}_p) \to R/R^p[F, R] \to H_1(F, \mathbf{Z}_p) \to H_1(G, \mathbf{Z}_p) \to 0.$$

Let K be the kernel of $H_1(F, \mathbf{Z}_p) \to H_1(G, \mathbf{Z}_p)$. Then we have the exact sequences

$$(5) \qquad 0 \to H_2(G, \mathbf{Z}_p) \to R/R^p[F, R] \to K \to 0$$

$$(6) \qquad 0 \to K \to H_1(F, \mathbf{Z}_p) \to H_1(G, \mathbf{Z}_p) \to 0.$$

Case $p = 0$. $H_1(F, \mathbf{Z}) = F/[F, F]$ is a free abelian group of rank $n + r$, and by the hypothesis the rank of $H_1(G, \mathbf{Z})$ is less than or equal to n. Hence, by the exactness of the sequence (6), K is free of rank greater than or equal to r. But since $R/[F, R]$ is generated by r elements, the exactness of the sequence (5) implies the rank of K is r. Hence in (6), K is of rank r and $H_1(F, \mathbf{Z})$ is of rank $n + r$, so $H_1(G, \mathbf{Z})$ is of rank n. This together with the hypothesis that $H_1(G, \mathbf{Z})$ is generated by n elements implies $H_1(G, \mathbf{Z})$ is free abelian freely generated by n elements. The group K is free abelian since it is a subgroup of the free abelian group $H_1(F, \mathbf{Z})$, and hence, the sequence (5) splits. This means that the free abelian group K of rank r is a subgroup of the group $R/[F, R]$ generated by r elements. Hence $K = R/[F, R]$ and so $H_2(G, \mathbf{Z}) = 0$. Let $\{g_1, \ldots, g_n\}$ be a subset of G whose image under the canonical map $G \to G/[G, G]$ is a free generating set. Let H be a free group freely generated by the symbols $h_1, \ldots, h_n$. Then the homomorphism $H \to G$ defined by $h_\iota \to g_\iota$ induces an isomorphism $H_1(H, \mathbf{Z}) \cong H_1(G, \mathbf{Z})$, and an epimorphism $H_2(H, \mathbf{Z}) \to H_2(G, \mathbf{Z})$ since both $H_2(H, \mathbf{Z})$ and $H_2(G, \mathbf{Z})$ are trivial. Hence, by 58.4 we have $H/H_m \cong G/G_m$. Moreover since, by 60.3, $\bigcap_{m=0}^{\infty} H_m$ is trivial, by 58.4, the homomorphism $H \to G$ is an injection. Thus 61.3 holds for $p = 0$.

Case $p \neq 0$. Since $R/R^p[F, R]$ is a factor group of $R/[F, R]$, it is generated by r elements. It follows, by the same arguments as in the case $p = 0$, that K is a $\mathbf{Z}_p$-vector space of dimension r, $H_1(G, \mathbf{Z}_p)$ is of dimension n, and $H_2(G, \mathbf{Z}_p) = 0$. Thus 61.3 also holds for $p \neq 0$.

61.3 in the case $p = 0$ implies the following result first obtained by Magnus [73].

61.4 *Let G be a group presented by $n + r$ generators and r relations. If G is also presented by n generators, then G is freely generated by n generators.*

PROOF Since $G/[G, G] = H_1(G, \mathbf{Z})$ is generated by the image of the generating set $g_1, \ldots, g_n$ of G under the canonical map $\theta: G \to G/[G, G]$, the hypothesis of 61.3 is satisfied in the case $p = 0$. Therefore, by 61.3, G is freely generated by the elements $g_1, \ldots, g_n$.

62. The Schur Multiplicator

In Chapter 4 we used the cohomology group $H^2(G, \mathbf{Q}/\mathbf{Z})$ to study dimension subgroups; $H^2(G, \mathbf{Q}/\mathbf{Z})$ is called the Schur multiplicator of G and in this section will be denoted by $M(G)$. The main result in this section is an estimate for the order of the Schur multiplicator of a finite nilpotent group. For a finite p-group G of order p^m Green [41] has shown that

$$|M(G)| \leqq p^{m(m-1)/2}.$$

For a finite p-group G another estimate,

$$|M(G)| \leqq |M(G/[G, G])| \cdot |[G, G]|^{d(G/Z(G))-1}, \tag{7}$$

was given by Gaschutz *et al.* [36] where $Z(G)$ is the center of G and $d(G)$ is the minimum number of generators of G. In this section we show that the inequality (7) holds if G is any finite nilpotent group. This result is due to Vermani [109].

Let G be a finite nilpotent group of class n. Let $G = G_0 > G_1 > \cdots > G_{n-1} > G_n = \{1\}$ be the lower central series of G. Then G_{n-1} is contained in the center of G. Consider the exact sequence

$$1 \to G_{n-1} \overset{i}{\to} G \underset{\rho}{\overset{j}{\rightleftarrows}} G/G_{n-1} \to 1$$

where ρ is a cross section, i.e., $\rho(1 \cdot G_{n-1}) = 1$ and $j\rho = 1_{G/G_{n-1}}$. Then the elements of G can be uniquely written in the form $\rho(h)z$ where $h \in G/G_{n-1}$ and $z \in G_{n-1}$.

62.1 (Passi [86]) *Let G be a nilpotent group of class n and let $\xi \in H^2(G, \mathbf{Q}/\mathbf{Z})$. Then there exists a 2-cocycle $f: G \times G \to \mathbf{Q}/\mathbf{Z}$ representing ξ such that*

(i) $f(z_1, z_2) = 0$ *for any* $z_1, z_2 \in G_{n-1}$,
(ii) $f(z, \rho(h)) = 0$ *for any* $z \in G_{n-1}$, $h \in G/G_{n-1}$.
(iii) $f(\rho(h_1)z_1, \rho(h_2)z_2) = f(\rho(h_1), \rho(h_2)) + f(\rho(h_1), z_2)$ *for any* $z_1, z_2 \in G_{n-1}$; $h_1, h_2 \in G/G_{n-1}$.

Proof Let $\varphi: G \times G \to \mathbf{Q}/\mathbf{Z}$ be a normalized 2-cocycle representing ξ. We claim that the restriction of φ to $G_{n-1} \times G_{n-1}$ is a symmetric 2-cocycle. For if $x, y \in G_{n-1}$, then $x - 1 \in I_G^n \subseteq I_G^2$ where I_G is the augmentation ideal of G. Extend φ by linearity to φ: $\mathbf{Z}G \times \mathbf{Z}G \to \mathbf{Q}/\mathbf{Z}$. Then $\varphi(x, y) = \varphi(x - 1, y)$; moreover it is enough to show $\varphi(x - 1, y) = \varphi(y, x - 1)$ in the case where $x - 1 = (g_1 - 1)(g_2 - 1)$. The verification that

$$\varphi((g_1 - 1)(g_2 - 1), y) = \varphi(y, (g_1 - 1)(g_2 - 1))$$

follows from the fact that φ is a 2-cocycle and y is in the center of G. Hence the extension

$$1 \to \mathbf{Q}/\mathbf{Z} \to E \to G_{n-1} \to 1$$

corresponding to the restriction of φ to $G_{n-1} \times G_{n-1}$ is abelian (this is easy to verify from the definition of a 2-cocycle). Moreover this extension splits since $\mathbf{Q}/\mathbf{Z}$ is injective in the category of abelian groups (Section 26). Hence φ represents the null element in $H^2(G_{n-1}, \mathbf{Q}/\mathbf{Z})$, i.e., there exists a 1-cochain $\chi\colon G_{n-1} \to \mathbf{Q}/\mathbf{Z}$ such that $\varphi(z_1, z_2) = \chi(z_2) - \chi(z_1 z_2) + \chi(z_1)$ for any $z_1, z_2 \in G_{n-1}$. Define $\psi\colon G \to \mathbf{Q}/\mathbf{Z}$ by $\psi(g) = \varphi(z, \rho(h) - \chi(z))$, where $g = \rho(h)z$, $z \in G_{n-1}$, $h \in G/G_{n-1}$. Clearly ψ is a 1-cochain, and so $f\colon G \times G \to \mathbf{Q}/\mathbf{Z}$ defined by

$$f(g_1, g_2) = \varphi(g_1, g_2) + \psi(g_1) - \psi(g_1 g_2) + \psi(g_2)$$

is a 2-cocycle. Moreover from the definitions it is immediate that

$$(8) \qquad f(z_1, z_2) = 0 \qquad \text{for} \quad z_1, z_2 \in G_{n-1}$$

$$(9) \qquad f(z, \rho(h)) = 0 \qquad \text{for} \quad z \in G_{n-1}, \quad h \in G/G_{n-1}.$$

This completes the proof of (i) and (ii). To show f also satisfies (iii), we first establish the identity

$$(10) \qquad f((\rho(h_1) - 1)(\rho(h_2) - 1), z) = 0 \qquad \text{for} \quad z \in G_{n-1}, \quad h_1, h_2 \in G/G_{n-1}.$$

$$\begin{aligned}
(11) \quad f((\rho(h_1) - 1)(\rho(h_2) - 1), z) &= f(z, (\rho(h_1) - 1)(\rho(h_2) - 1)) \\
&\qquad \text{since} \quad z \text{ is central} \\
&= f(z, \rho(h_1)\rho(h_2)) \qquad \text{by (8)} \\
&= f(z, \rho(h_1 h_2)z') \qquad \text{for some} \quad z' \in G_{n-1} \\
&= f(z, (\rho(h_1 h_2) - 1)(z' - 1) + \rho(h_1, h_2) + z') \\
&= f((z - 1)(z' - 1), \rho(h_1, h_2)) \qquad \text{by (8), (9),} \\
&\qquad \text{and the fact that } z' \text{ is central} \\
&= 0.
\end{aligned}$$

Then

$$\begin{aligned}
&f(\rho(h_1)z_1, \rho(h_2)z_2)\\
&\quad = f((\rho(h_1) - 1)(z_1 - 1) + \rho(h_1) + z_1, (\rho(h_2) - 1)(z_2 - 1)\\
&\qquad + \rho(h_2) + z_2)\\
&\quad = f((\rho(h_1) - 1)(z_1 - 1), (\rho(h_2) - 1)(z_2 - 1))\\
&\qquad + f((\rho(h_1) - 1)(z_1 - 1), \rho(h_2)) + f((\rho(h_1) - 1)(z_1 - 1), z_2)\\
&\qquad + f(\rho(h_1), (\rho(h_2) - 1)(z_2 - 1)) + f(\rho(h_1), \rho(h_2))\\
&\qquad + f(\rho(h_1), z_2) + f(z_1, (\rho(h_2) - 1)(z_2 - 1))\\
&\qquad + f(z_1, \rho(h_2)) + f(z_1, z_2),
\end{aligned}$$

where the first term on the right-hand side is equal to $f((\rho(h_1) - 1)(\rho(h_2 - 1)), (z_1 - 1)(z_2 - 1))$ and is zero by (10). Similarly the second and fourth terms on the right-hand side are zero by (10). The third term on the right-hand side is equal to

$$f(\rho(h_1), (z_1 - 1)(z_2 - 1)) = f((z_1 - 1)(z_2 - 1), \rho(h_1))$$

and is zero by (9). Similarly the seventh and eighth terms on the right-hand side are zero by (9). Finally the ninth term is zero by (8). Therefore

$$f(\rho(h_1)z_1, \rho(h_2)z_2) = f(\rho(h_1), \rho(h_2)) + f(\rho(h_1), z_2),$$

which is the required identity.

62.2 (Vermani [109]) *Let G be a nilpotent group of class n. Let f, f_1, f_2 be normalized* 1-*cocycles which satisfy the identity 62.1,* (iii). *Then*

(i) $f_1(g, z) = f_2(g, z)$, $g \in G$, $z \in G_n$ *if f_1 and f_2 represent the same element $\xi \in H^2(G, \mathbf{Q}/\mathbf{Z})$.*

(ii) *The restriction of f to $G \times G_{n-1}$ is bilinear.*

(iii) *If $\{1\} = Z_0(G) < Z_2(G) < \cdots < Z_{n-1}(G) < Z_n(G) = G$ is the upper central series of G, then the restriction of f to $G_{n-1} \times Z_{n-1}(G)$ is the zero map.*

Proof of (i) $f_1 - f_2$ is a 2-coboundary, therefore there exists a 1-cochain $u\colon G \to \mathbf{Q}/\mathbf{Z}$ such that $f_1 - f_2 = \delta u$. In particular, for $g \in G$ and $z \in G_{n-1}$,

$$
\begin{aligned}
(f_1 - f_2)(g, z) &= u(z) - u(gz) + u(g) \\
&= u(z) - u(zg) + u(g) \\
&= (f_1 - f_2)(z, g) \\
&= 0.
\end{aligned}
$$

Thus $f_1(g, z) = f_2(g, z)$.

PROOF OF (ii) Since f is a cocycle,

$$
\begin{aligned}
f(g, z_1 z_2) &= -f(z_1, z_2) + f(gz_1, z_2) + f(g, z_1) \\
&= f(g, z_2) + f(g, z_1) \qquad \text{by 62.1, (iii).}
\end{aligned}
$$

Similarly

$$
\begin{aligned}
f(g_1 g_2, z) &= f(g_2, z) + f(g_1, g_2 z) - f(g_1, g_2) \\
&= f(g_2, z) + f(g_1, z) \qquad \text{by 62.1, (iii).}
\end{aligned}
$$

The reader may fill in the detail above by observing that 62.1, (iii) can be used to show that $f(gz_1, z_2) = f(g, z_2)$ and $f(g_1, g_2 z) = f(g_1, g_2) + f(g_1, z)$.

PROOF (iii) Let

$$
1 \to \mathbf{Q}/\mathbf{Z} \xrightarrow{i} H \underset{\rho}{\overset{j}{\rightleftarrows}} G \to 1
$$

be the extension corresponding to the cocycle class of f, and the cross section corresponding to the cocycle f. Since $\mathbf{Q}/\mathbf{Z}$ is in the center of H, the class of H is less than or equal to $n + 1$. We claim $g \in Z_{n-1}(G)$ implies $\rho(g) \in Z_n(H)$. For, let $h_1, \ldots, h_{n-1} \in H$, then

$$
j([\cdots[[\rho(g), h_1], h_2], \ldots, h_{n-1}]) = [\cdots[[g, jh_1], jh_2], \ldots, jh_{n-1}].
$$

Hence $[\cdots[[\rho(g), h_1], h_2], \ldots, h_{n-1}] \in \mathbf{Q}/\mathbf{Z} \subseteq Z(H)$. Therefore $[[\cdots[[\rho(g), h_1], h_2], \ldots, h_{n-1}], h_n] = 1$ for any $h_1, \ldots, h_n \in H$. This implies $\rho(g) \in Z_n(H)$ since H is of class less than or equal to $n + 1$. Since $i(\mathbf{Q}/\mathbf{Z})$ is central in H, $\rho(z) \in H_{n-1} \cdot \mathbf{Q}/\mathbf{Z}$ for every $z \in G_{n-1}$. Also $[Z_n(H), H_{n-1}] \leqq Z_0(H) = \{1\}$ since in general $[Z_r(H), H_{s-1}] \leqq Z_{r-s}(H)$ whenever $r \geqq s$. It follows from these facts that if $g \in Z_{n-1}(G)$ and $z \in G_{n-1}$ we have $\rho(g)\rho(z) = \rho(z)\rho(g)$. Hence $f(g, z) = f(z, g) = 0$.

62.3 *Let G be a nilpotent group of class $n > 1$. Then there exists a homomorphism α such that the sequence*

$$(12) \qquad H^2(G/G_{n-1}, \mathbf{Q}/\mathbf{Z}) \xrightarrow{\text{inf}} H^2(G, \mathbf{Q}/\mathbf{Z}) \xrightarrow{\alpha} \operatorname{Hom}((G/Z_{n-1}(G)) \otimes G_{n-1}, \mathbf{Q}/\mathbf{Z})$$

is exact.

Proof To define α, let $\xi \in H^2(G, \mathbf{Q}/\mathbf{Z})$ and f be a 2-cocycle representing ξ which satisfies 62.1, (iii). Define $\eta: G \times G_{n-1} \to \mathbf{Q}/\mathbf{Z}$ by $\eta(g, z) = f(g, z)$. By 62.2, (i), η is independent of the choice of f. By 62.2, (ii)–(iii), η is bilinear and induces a bilinear map $(G/Z_{n-1}(G)) \times G_{n-1} \to \mathbf{Q}/\mathbf{Z}$. Therefore η induces a unique homomorphism

$$\bar{\eta}: (G/Z_{n-1}(G)) \otimes G_{n-1} \to \mathbf{Q}/\mathbf{Z}.$$

Define

$$\alpha(\xi) = \bar{\eta}.$$

α is a homomorphism. To show $\alpha \inf = 0$, let $\xi \in H^2(G/G_{n-1}, \mathbf{Q}/\mathbf{Z})$ and f be a 2-cocycle representing ξ. Then $\inf(\xi)$ is the element of $H^2(G, \mathbf{Q}/\mathbf{Z})$ represented by the 2-cocycle

$$f'(g_1, g_2) = f(g_1 G_{n-1}, g_2 G_{n-1}).$$

We observe that f' satisfies 62.1, (iii). Hence, for $gZ_{n-1}(G) \in G/Z_{n-1}(G)$, $z \in G_{n-1}$,

$$\alpha(\inf(\xi)) = f'(g, z) = 0.$$

Thus (12) is a complex. To show exactness at $H^2(G, \mathbf{Q}/\mathbf{Z})$, let $\xi \in \ker \alpha$; i.e., $\alpha(\xi)(gZ_{n-1}(G) \otimes z) = 0$ for any $g \in G$, $z \in G_n$. Let f be an element in the cocycle class of ξ which satisfies 62.1, (iii). Then

$$f(g, z) = \alpha(\xi)(gZ_{n-1}(G) \otimes z) = 0.$$

Hence

$$f(g_1, g_2) = f(\rho(h_1), \rho(h_2)).$$

Define $f': G/G_{n-1} \times G/G_{n-1} \to \mathbf{Q}/\mathbf{Z}$ by $f'(h_1, h_2) = f(\rho(h_1), \rho(h_2))$; f' is a 2-cocycle. Let ξ' be the cocycle class of f', then $\inf(\xi') = \xi$. Thus the complex (12) is exact at $H^2(G, \mathbf{Q}/\mathbf{Z})$.

62.4 *Let G be a nilpotent group of class $n > 1$. Then the sequence*

$$0 \to \mathrm{Hom}(G_{n-1}, \mathbf{Q}/\mathbf{Z}) \xrightarrow{t} H^2(G/G_{n-1}, \mathbf{Q}/\mathbf{Z}) \xrightarrow{\inf} H^2(G, \mathbf{Q}/\mathbf{Z})$$

$$\xrightarrow{\alpha} \mathrm{Hom}((G/Z_{n-1}(G) \otimes G_{n-1}, \mathbf{Q}/\mathbf{Z}),$$

where t is the transgression map, is exact.

PROOF Follows from the five-term Hochschild–Serre exact sequence and the sequence (12) in 62.3, the fact that the restriction map $\mathrm{Hom}(G, \mathbf{Q}/\mathbf{Z}) \to \mathrm{Hom}(G_{n-1}, \mathbf{Q}/\mathbf{Z})$ is the zero map, and the identity $H^1(G, \mathbf{Q}/\mathbf{Z}) = \mathrm{Hom}(G, \mathbf{Q}/\mathbf{Z})$ since the action of G on $\mathbf{Q}/\mathbf{Z}$ is trivial.

62.5 (Vermani [109]) *Let $M(G)$ be the Schur multiplicator of the group G. If G is a finite nilpotent group*

$$|M(G)| \leqq |M(G/[G, G])| \cdot |[G, G]|^{d(G/Z(G))-1}, \tag{13}$$

where $d(G/Z(G))$ is the minimum number of generators of G modulo its center.

PROOF Since $G/Z_{n-1}(G)$, G_{n-1}, and $\mathbf{Q}/\mathbf{Z}$ are abelian groups,

$$\mathrm{Hom}((G/Z_{n-1}(G)) \otimes G_{n-1}, \mathbf{Q}/\mathbf{Z}) = \mathrm{Hom}(G/Z_{n-1}(G), \mathrm{Hom}(G_{n-1}, \mathbf{Q}/\mathbf{Z})),$$

by 47.1. Moreover since G_{n-1} is finite $\mathrm{Hom}(G_{n-1}, \mathbf{Q}/\mathbf{Z}) \cong G_{n-1}$. Hence the exactness of the sequence in 62.4 implies

$$|M(G)| \leqq \frac{|M(G/G_{n-1})| \cdot |\mathrm{Hom}(G/Z_{n-1}(G), G_{n-1})|}{|G_{n-1}|}. \tag{14}$$

To get the required estimate (13) for $|M(G)|$ from (14) we proceed by induction on the class of G. Since (13) is trivially true for groups of class 1, suppose (13) true for groups of class less than n. Then, observing that $[G/G_{n-1}, G/G_{n-1}] = [G, G]/G_{n-1}$, we have

$$|M(G/G_{n-1})| \leqq |M(G/[G, G])| \cdot |[G, G]/G_{n-1}|^{\delta-1}, \tag{15}$$

where $\delta = d((G/G_{n-1})/Z(G/G_{n-1}))$. Since $Z(G)/G_{n-1} \leqq Z(G/G_{n-1})$, we have $\delta \leqq d(G/Z(G))$. This together with (14) and (15) implies

$$(16)\qquad \begin{aligned} |G_{n-1}|\cdot|M(G)| &\leqq |M(G/G_{n-1})|\cdot|\mathrm{Hom}(G/Z_{n-1}(G), G_{n-1})| \\ &\leqq |M(G/[G, G])|\cdot|[G, G]/G_{n-1}|^{d(G/Z(G))-1} \\ &\quad\cdot|\mathrm{Hom}(G/Z_{n-1}(G), G_{n-1})|. \end{aligned}$$

Since $|\mathrm{Hom}(G/Z_{n-1}(G), G_{n-1})| \leqq |G_{n-1}|^{d(G/Z_{n-1}(G))} \leqq |G_{n-1}|^{d(G/Z(G))}$, (16) implies

$$|M(G)| \leqq |M(G/[G, G])|\cdot|[G, G]|^{d(G/Z(G))-1}.$$

CHAPTER

10

Galois Groups

This chapter contains a brief account of the use of cohomology groups in Galois theory.

63. Galois Groups

63.1 Let K, F be fields, and $\sigma_1, \sigma_2, \ldots, \sigma_n$ monomorphisms of K into F. Then $\sigma_1, \sigma_2, \ldots, \sigma_n$ are called *linearly independent* if the relation

$$c_1\sigma_1(x) + \cdots + c_n\sigma_n(x) = 0 \qquad \text{for every} \quad x \in K,$$

where $c_i \in F$, implies $c_1 = c_2 = \cdots = c_n = 0$.

If $\sigma_1, \sigma_2, \ldots, \sigma_n$ are distinct, then they are linearly independent.

PROOF The proof is by induction on n. If $n = 1$, $c_1\sigma_1(x) = 0$ does imply $c_1 = 0$. Let $m \leqq n$ be the smallest number such that after renumbering of the σ_i,

$$(1) \qquad c_1\sigma_1(x) + \cdots + c_m\sigma_m(x) = 0 \qquad \text{for every} \quad x \in K,$$

with not all $c_i = 0$. Choose $a \in K$ such that $\sigma_1(a) \neq \sigma_m(a)$. Replace x by ax in (1) to get

$$c_1\sigma_1(a)\sigma_1(x) + \cdots + c_m\sigma_m(a)\sigma_m(x) = 0 \qquad \text{for every} \quad x \in K. \tag{2}$$

Next multiply (1) by $\sigma_m(a)$ to get

$$c_1\sigma_m(a)\sigma_1(x) + \cdots + c_m\sigma_m(a)\sigma_m(x) = 0 \qquad \text{for every} \quad x \in K. \tag{3}$$

Subtract (2) from (3) to get a relation of linear dependence for $\sigma_1, \ldots, \sigma_{m-1}$, contradicting the minimality of m.

REMARK We have shown, in particular, that there exists $\alpha \in K$ such that $\sum_{i=1}^{n} \sigma_i(\alpha) \neq 0$.

63.2 *Let k be the set of all elements α in K such that if $\sigma_1, \ldots, \sigma_n$ are distinct monomorphisms of K into F, then $\sigma_1(\alpha) = \sigma_2(\alpha) = \cdots = \sigma_n(\alpha)$; k is clearly a field. We contend $[K\colon k] \geqq n$.*

Proof Suppose $\{\omega_1, \ldots, \omega_m\}$ is a basis of K over k with $m < n$. Then the system of equations

$$\sum_{i=1}^{n} X_i\sigma_i(\omega_j) = 0, \qquad j = 1, \ldots, m \tag{4}$$

has a nontrivial solution $(c_1, \ldots, c_n) \in F^n$. Since any $x \in K$ can be written $x = x_1\omega_1 + \cdots + x_n\omega_n$, substitute the solution $(c_1, \ldots, c_n)$ in (4), multiply (4) by x_j, and sum over j to get

$$c_1\sigma_1(x) + \cdots + c_n\sigma_n(x) = 0 \qquad \text{for any} \quad x \in K.$$

This contradicts the linear independence of $\sigma_1, \ldots, \sigma_n$.

63.3 *Let $\sigma_1, \ldots, \sigma_n$ be distinct automorphisms of K. Suppose the identity automorphism is among them. Let k be as in 63.2. Then $[K\colon k] = n$ if and only if $\sigma_1, \ldots, \sigma_n$ form a group.*

PROOF Suppose $\sigma_1, \ldots, \sigma_n$ form a group. Let $\alpha_1, \ldots, \alpha_{n+1} \in K$; we will show they are linearly dependent over k. Solve the system of n equations with $n + 1$ unknowns

$$\sum_{i=1}^{n+1} X_i\sigma_j^{-1}(\alpha_i) = 0, \qquad j = 1, \ldots, n \tag{5}$$

to get a nontrivial solution $(c_1, \ldots, c_{n+1}) \in K^{n+1}$. Say $c_1 \neq 0$. Since the system is homogeneous any multiple of $(c_1, \ldots, c_{n+1})$ by an element of K is again a solution. Choose a solution such that the trace of c_1, $T(c_1) \neq 0$ (this is possible by the remark at the end of 63.1). Substitute this solution in (5), multiply (5) by σ_j, and sum over j to get

$$T(c_1)\alpha_1 + T(c_2)\alpha_2 + \cdots + T(c_{n+1})\alpha_{n+1} = 0.$$

Since $T(c_i) \in k$ for all i and $T(c_1) \neq 0$, this is a relation of linear dependence for $\alpha_1, \ldots, \alpha_{n+1}$ over k. Hence $[K: k] \leqq n$, so, by 63.2, $[K: k] = n$. Conversely, since, by 63.2, the set of distinct automorphisms cannot exceed $[K: k]$, the fixed field k cannot change by adjoining the inverses and products of $\sigma_1, \ldots, \sigma_n$ to this set.

The group $\sigma_1, \ldots, \sigma_n$ is called the *Galois group* of K over k.

63.4 The extension K of k is called (separably) *normal* if k is the fixed field of some finite group of automorphisms of K.

If $k \subset F \subset K$, and K/k (K over k) is normal, then so is K/F.

Proof Let G be the Galois group of K/k and $U = \{\sigma \in G, \text{ s.t. } \sigma(a) = a \text{ for every } a \in F\}$. Then $U \leqq G$. Let $\sigma_1, \sigma_2 \in G$. Then res σ_1 to F = res σ_2 to F if and only if $\sigma_2 \in \sigma_1 U$. Hence the number of distinct restrictions to F is $(G: U)$. By 63.2, $[F: k] \geqq (G: U)$. Let F' be the fixed seld of U. Then (1) $F \subset F'$; (2) $[K: F'] \geqq (U: 1)$. If $[K: k] = n$, we have

$$[F': k] = n/[K: F'] \leq n/(U: 1) = (G: U) \leq [F: k].$$

Comparing with (1) we conclude $F = F'$.

63.5 (i) *The subgroup U of G is the Galois group of K/F.*
(ii) *F/k is normal if and only if U is normal in G.*
(iii) *If F/k is normal, then G/U is its Galois group.*

Proof The first statement is a consequence of 63.4. To show (ii) and (iii) observe that the number of left cosets of U in G is $(G: U)$. Next observe that the restrictions of all $\sigma \in G$ induce $(G: U)$ distinct classes of isomorphisms of F into K. If F is normal over k, these $(G: U)$ isomorphisms are automorphisms, i.e., $\sigma \in G$, $\sigma|_F(F) = F$; so $\sigma U \sigma^{-1} = U$, thus $U \trianglelefteq G$. Conversely if $\sigma U \sigma^{-1} = U$ for all $\sigma \in G$, then $\sigma F = F$. Thus the $(G: U)$ distinct classes of isomorphisms of F into K are classes of automorphisms of F; moreover the fixed elements $\{x\}$ of F under these automorphisms satisfy $\sigma x = x$ for all $\sigma \in G$ and so must be the elements of k. Hence F/k is normal.

63.6 Let K/k be normal. Then K/k is separable. *Moreover any element of K is a root of a polynomial over k which splits completely in K.*

PROOF Let $G = \{\sigma_1, \ldots, \sigma_n\}$ be the Galois group of K/k. Let $\alpha \in K$. Compute $\sigma_1\alpha, \ldots, \sigma_n\alpha$ and after a suitable permutation of indices let $\sigma_1\alpha, \ldots, \sigma_r\alpha$ be a maximal subset of distinct elements of the set $\{\sigma_1\alpha, \ldots, \sigma_n\alpha\}$. Let

$$f(t) = \prod_{i=1}^{r} (t - \sigma_i(\alpha)) \in K[t].$$

Since $\sigma_j f(t) = f(t)$, we conclude that $f(t) \in k[t]$. Conversely, let $g(t) \in k[t]$ be of minimal degree such that $g(\alpha) = 0$. Then $\sigma_j g(\alpha) = 0$ for all j. Hence $\deg g(t) \geqq \deg f(t)$, and so $f(t)$ is irreducible.

63.7 *Consider the situation*

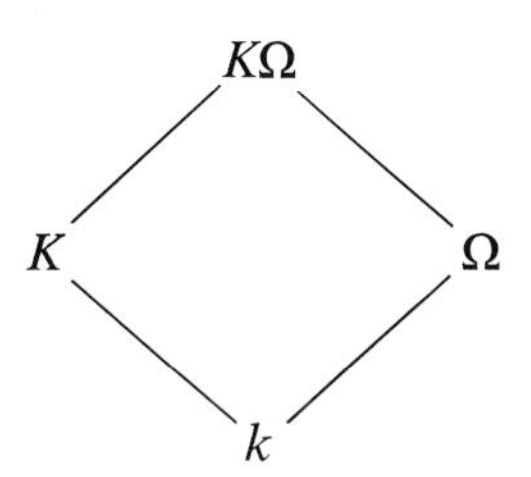

where K/k is normal with Galois group G; Ω is an extension of k. Both K and Ω are subfields of some extension of k. Then

(i) *$K\Omega/\Omega$ is normal; if K/k is the splitting field of $f(t)$, then $K\Omega/\Omega$ is the splitting field of $f(t)$.*

(ii) *Let U be the Galois group of $K\Omega/\Omega$; then the map $\rho: U \to G$ defined by $\rho(v) = v|_K$ is a monomorphism. Moreover $\rho(U)$ is the Galois group of $K/K \cap \Omega$.*

PROOF $\rho(v) \in G$ since K/k is normal. If $\rho(v) = 1$ then $v = 1 \in U$ clearly; $\rho(U)$ leaves fixed those elements of $K\Omega$ which U leaves fixed and which lie in K. Therefore the fixed field of $\rho(U)$ is $K \cap \Omega$.

63.8 Consider the situation

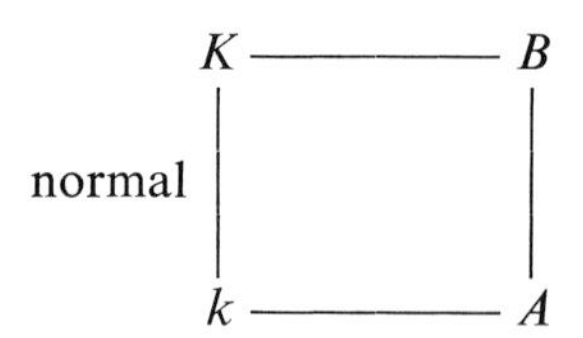

where K/k is normal with Galois group G, A is a subring of k, B is a subring of K with quotient field K. Assume

(i) A is integrally closed in k.
(ii) B/A is integral (so $B \cap k = A$).
(iii) B is invariant under G (i.e., $\sigma B = B$).

Let $\bar{B}$ be an integral domain such that there exists an epimorphism $\varphi: B \to \bar{B}$. Let $\bar{A}$ be the image of restriction of φ to A. Let $\bar{K}$ be the quotient field of $\bar{B}$ and $\bar{k}$ the quotient field of $\bar{A}$. Diagrammatically we have

$$
\begin{array}{ccccccc}
K & \supset & B & \xrightarrow{\varphi} & \bar{B} & \subset & \bar{K} \\
| & & | & & | & & | \\
k & \supset & A & \xrightarrow{\varphi|_A} & \bar{A} & \subset & \bar{k}.
\end{array} \tag{6}
$$

Assume further that $\bar{K}/\bar{k}$ is finite separable normal with Galois group $\bar{G}$. We wish to study the relation between G and $\bar{G}$. For $\alpha \in B$ denote $\varphi(\alpha) = \bar{\alpha}$. Let $F(t) = \prod_{\sigma \in G} (t - \sigma\alpha)$. Then the coefficients of $F(t)$ belong to B. But the coefficients of $F(t)$ are invariant under σ, so they are in k. But $B \cap k = A$, therefore the coefficients of $F(t)$ are in A. Since $F(\alpha) = 0$, so is $\varphi F(\alpha) = \bar{F}(\bar{\alpha}) = 0$ ($\bar{F}$ is obtained by applying φ to the coefficients of F). This is a polynomial relation from $\bar{\alpha}$ over $\bar{k}$. Let $\zeta \in \bar{G}$. Then $\bar{F}(\zeta\bar{\alpha}) = 0$ because the coefficients of $\bar{F}(t)$ are in $\bar{A}$. Since

$$\varphi F(t) = \bar{F}(t) = \prod_{\sigma \in G} (t - \bar{\sigma}\bar{\alpha}),$$

there exists $\sigma \in G$ such that $\zeta\bar{\alpha} = \bar{\sigma}\bar{\alpha}$. This selection of σ depends on the choice of α. We contend there is a $\sigma \in G$ such that $\zeta\bar{\alpha} = \bar{\sigma}\bar{\alpha}$ for every $\alpha \in B$. Let x be transcendental over K. Consider the diagram

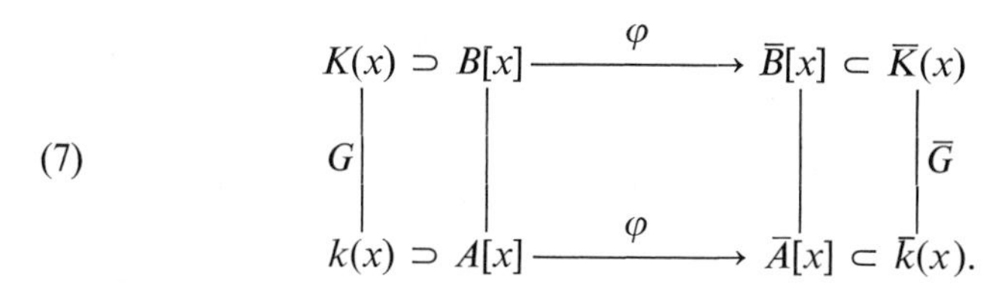

All the assumptions of the diagram (6) also hold here (the proof is left to the reader). Observe in particular that G (resp. $\bar{G}$) is the Galois group of $K(x)/k(x)$ [resp. $\bar{K}(x)/\bar{k}(x)$], and φ is applied coefficientwise in (7). Let $\alpha_0, \ldots, \alpha_m \in B$. Let $\xi = \alpha_0 + \alpha_1 x + \cdots + \alpha_m x^m \in B[x]$, then $\varphi\xi = \bar{\xi} = \bar{\alpha}_0 + \bar{\alpha}_0 x + \cdots + \bar{\alpha}_m x^m$. Repeating the procedure applied to the diagram (6), for $\zeta \in \bar{G}$ there exists $\sigma \in G$ such that $\zeta\bar{\xi} = \bar{\sigma}\bar{\xi}$. Hence

$$\zeta\bar{\alpha}_0 + \zeta\bar{\alpha}_1 x + \cdots + \zeta\bar{\alpha}_m x^m = \bar{\sigma}\bar{\alpha}_0 + \bar{\sigma}\bar{\alpha}_1 x + \cdots + \bar{\sigma}\bar{\alpha}_m x^m.$$

Comparing coefficients we conclude that for $\zeta \in \bar{G}$ and a finite number of α_i in B there is a $\sigma \in G$ such that $\zeta\bar{\alpha}_i = \bar{\sigma}\bar{\alpha}_i$. If we consider two such finite sets of α's, the procedure above will give a single σ for both. Thus any infinite sequence of such sets of α's will give an infinite sequence of σ's all taken from the finite group G. This implies that the sequence of the σ's so obtained has a point of accumulation, say $\sigma \in G$ (simple compactness argument). It follows by set theory that for every $\zeta \in \bar{G}$ there exists a $\sigma \in G$ such that $\zeta\bar{\alpha} = \bar{\sigma}\bar{\alpha}$ for every $\alpha \in B$. This, moreover, implies $\zeta\bar{B} = \bar{B}$ for every $\zeta \in \bar{G}$. Returning to the diagram (6), let P be the kernel of φ. Let $\zeta \in \bar{G}$; there exists $\sigma \in G$ such that $\zeta\alpha = \bar{\sigma}\bar{\alpha}$ for every $\alpha \in B$. This implies $\zeta\bar{\alpha} = 0 = \bar{\sigma}\bar{\alpha}$ for every $\alpha \in P$. Hence $\sigma P \subset P$. This means $\sigma^{n-1}P \subset P$ where $n = |G|$. But $\sigma^{n-1} = \sigma^{-1}$, therefore $\sigma^{-1}P \subset P$. Applying σ we get $P \subset \sigma P$. So $\sigma P = P$. Let $D = \{\sigma \in G, \text{ s.t. } \sigma P = P\}$; D is called the *decomposition group*. Hence the σ corresponding to ζ obtained above belongs to D. Next, let $\sigma \in D$. Define $\bar{\sigma}\bar{\alpha} = \overline{\sigma\alpha}$ for every $\bar{\sigma} \in \bar{B}$; $\bar{\sigma}$ is well defined, for, if $\bar{\alpha} = \bar{\beta}$, then $\alpha - \beta \in P$, so $\sigma\alpha - \sigma\beta \in P$. Hence $\overline{\sigma\alpha} = \overline{\sigma\beta}$. Moreover $\bar{\sigma}\bar{\tau}\bar{\alpha} = \bar{\sigma}\overline{\tau\alpha} = \overline{\sigma\tau\alpha} = \overline{\sigma\tau}\bar{\alpha}$ and $\bar{1} = 1$, $\overline{\sigma^{-1}} = \bar{\sigma}^{-1}$. Hence $\bar{\sigma}\colon \bar{B} \to \bar{B}$ is an isomorphism, and leaves $\bar{A}$ fixed since σ leaves A fixed. Furthermore $\bar{\sigma}$ can be extended to a $\bar{k}$-automorphism of $\bar{K}$. Thus $\bar{\sigma} \in \bar{G}$. Summarizing, we have shown there exists an epimorphism $D \to \bar{G}$. Let V be the kernel of this epimorphism; V is called the *ramification group*, and $V = \{\sigma \in G, \text{ s.t. } \sigma\alpha - \alpha \in P\}$. So we have shown any prime ideal $P \subseteq B$ gives rise to subgroups D and V of G and

$$D/V \cong \bar{G}.$$

63.9 Let the normal extension K/k of 63.8 be the splitting field of a monic polynomial $f(x) \in A[x]$. Then $f(x) = \prod_{i=1}^{n} (x - \alpha_i)$ in $K[x]$. Let $B = A[\alpha_1, \ldots, \alpha_n]$. As in 63.8 assume A is integrally closed. Let the epimorphism $\varphi: A \to \bar{A}$ be given. Write $\varphi f(x) = \bar{f}(x)$. We require that $\bar{f}(x)$ have no multiple roots; φ can be extended to an epimorphism $\varphi: B \to \bar{B}$ where the quotient field of $\bar{B}$ is the splitting field of $\bar{f}(x)$. Then $\bar{K}/\bar{k}$ is normal with Galois group $\bar{G}$. We claim in this setup, the ramification group $V = 1$. In fact, let $\sigma \in V$, then $\sigma\alpha - \alpha \in P$. In particular $\overline{\sigma\alpha_i} = \bar{\alpha}_i$ for $i = 1, \ldots, n$. So the roots of $\bar{f}(x)$ are equal, since $\overline{\sigma\alpha_i}$, α_i are roots of $\bar{f}(x)$. Hence $\bar{\alpha}_i = \bar{\alpha}_j$, and so $\alpha_i = \alpha_j$ which means $\sigma\alpha_i = \alpha_i$. This means σ leaves the elements of K fixed, hence $\sigma = 1$. We conclude that in this case $D \cong \bar{G}$.

In particular if $\bar{G}$ is cyclic generated by ζ (this is what happens in the applications), ζ permutes $\bar{\alpha}_1, \ldots, \bar{\alpha}_n$ as follows: $\bar{f}(x)$ may not be irreducible. Let $\Phi(x)$ be an irreducible factor of $\bar{f}(x)$; say $\Phi(x) = (x - \bar{\alpha}_1) \cdots (x - \bar{\alpha}_r)$. Then $\bar{k}(\bar{\alpha}_1) \cong \bar{k}(\bar{\alpha}_i)$, $i \leqq r$, since both $\bar{\alpha}_1$ and $\bar{\alpha}_i$ are roots of the same irreducible polynomial. This isomorphism is induced by some element $\zeta^{\nu} \in \bar{G}$. Therefore $\bar{G}$ can be characterized by the irreducible factors of $\bar{f}(x)$. This gives information about the subgroups of G because of the isomorphism $D \cong \bar{G}$.

Example Let $k = Q$ the rationals, $A = \mathbf{Z}$ the integers, and $f(x) = x^5 - x - 1$ in our discussion. The Galois group of $f(x)$ is isomorphic to a subgroup of the symmetric group S_5. Let φ be $\varphi: \mathbf{Z} \to \mathbf{Z}/p\mathbf{Z}$. When $p = 5$, $f(x)$ is irreducible modulo 5, hence its Galois group $\bar{G}$ is isomorphic to the subgroup of S_5 generated by the cycle (12345). When $p = 2$, $f(x)$ is reducible and we have

$$x^5 - x - 1 = (x^2 + x + 1)(x^3 + x^2 + 1) \pmod 2$$

where $x^2 + x + 1$ and $x^3 + x^2 + 1$ are irreducible modulo 2. Hence in this case $\bar{G}$ contains the permutation (12)(345), so also the cube of this, i.e., (12). Hence the Galois group G of $f(x)$ over the rationals has a 5-cycle and a 2-cycle. This implies $G \cong S_5$.

63.10 Let K/k be a finite separable normal extension with Galois group G. If K^+ is the additive group of K in 15.7, we have shown that $H^r(G, K^+) = 0$ for all $r \in Z$. Another G-module is $K^\times$, the multiplicative group of K.

We have

$$H^1(G, K^\times) = 0.$$

PROOF Let X_* be the standard complete resolution of G. We will write the composition in the G-module $K^\times$ multiplicatively. Let $f \in \operatorname{Hom}_G(X_1, K^\times)$ be a cocycle. Then

$$(\delta f)(\sigma, \tau) = f(\tau)^\sigma f(\sigma\tau)^{-1} f(\sigma) = 1.$$

Let $\theta \in K^\times$, then

$$\sum_{\tau \in G} \theta^{\sigma\tau} f(\sigma\tau) = \sum_{\tau \in G} \theta^{\sigma\tau} f(\sigma) f(\tau)^\sigma \tag{8}$$

$$= f(\sigma) \sum_{\tau \in G} \theta^{\sigma\tau} f(\tau)^\sigma$$

$$= f(\sigma) (\sum_{\tau \in G} \theta^\tau f(\tau))^\sigma.$$

There exists $\theta \in K^\times$ such that

$$\prod_{\tau \in G} \theta^\tau f(\tau) \neq 0.$$

[The nonexistence of such θ would mean the automorphisms $\tau \in G$ are linearly dependent (63.1).] For such θ, denote $\alpha^{-1} = \sum_{\tau \in G} \theta^\tau f(\tau)$. Then, by (8),

$$\alpha^{-1} = f(\sigma)\alpha^{-\sigma}.$$

Thus $f(\sigma) = \alpha^{\sigma - 1}$, i.e., f is a 1-coboundary.

63.11 If G is cyclic, its cohomology groups are periodic of period 2. Hence $H^r(G, K^\times) = 0$ for odd r. In particular $H^{-1}(G, K^\times) = 1$ (written multiplicatively) means a 1-cocycle α satisfies the identity

$$\alpha^{1 + \sigma + \cdots \sigma^{n-1}} = 1,$$

i.e., norm of α equals 1. This statement was first recognized by Hilbert.

Hilbert's Theorem 90. *Let K/k be a finite normal separable extension with cyclic Galois group G. Let σ be a generator of G. Then any $\alpha \in K^\times$ which satisfies*

$$\alpha^{1 + \sigma + \cdots + \sigma^{n-1}} = 1$$

can be written as $\alpha = \beta^{\sigma - 1}$ *for some* $\beta \in K$.

PROOF Define the 1-cochain f by

$$f(\sigma^i) = \alpha^{1 + \sigma + \cdots + \sigma^{i-1}}.$$

f is well defined, for, $f(\sigma^{i+n}) = f(\sigma^i)(\alpha^{(1 + \sigma + \cdots + \sigma^{n-1})})^{\sigma^i} = f(\sigma^i)$; f is a cocycle since it is a crossed homomorphism, i.e.,

$$f(\sigma^i \cdot \sigma^j) = f(\sigma^i)(\alpha^{1 + \sigma + \cdots + \sigma^{j-1}})^{\sigma^i} = f(\sigma^i)f(\sigma^j)^{\sigma^i}.$$

Since 1-cocycles are 1-coboundaries, there exists $\beta \in K^\times$ such that $f(\sigma^i) = \beta^{\sigma^i - 1}$. For $i = 1$ we have

$$\alpha = \beta^{\sigma - 1}.$$

64. Abelian Galois Groups

64.1 Let G be a finite abelian group (written additively). Let $\hat{G} = \mathrm{Hom}_{\mathbf{Z}}(G, \mathbf{Q}/\mathbf{Z})$, the group of characters of G. We have seen in 47.7 that a character of a subgroup U of G can be extended to a charcter of G.

If G/U *is cyclic, a character of* U *can be extended to a character of* G *in* $(G: U)$ *ways.*

Write $G = \dot{\bigcup}_{i=0}^{e-1} (i\sigma + U)$ where $e = (G: U)$. Let $\chi: U \to \mathbf{Q}/\mathbf{Z}$ be given, and $\chi': G \to \mathbf{Q}/\mathbf{Z}$ the extension of χ constructed in 47.7; i.e., if $\sigma + U$ is a generator of G/U, and $e\sigma = \tau \in U$, set $\chi'(\sigma) = (1/e)\chi(\tau)$. Then for any element $i\sigma + \nu \in G$

$$\chi'(i\sigma + \nu) = \frac{1}{e}\chi(\tau) + \chi(\nu).$$

Now for $\eta \in \{1, 2, \ldots, e - 1\}$ define $\chi'_\eta: G \to \mathbf{Q}/\mathbf{Z}$ by $\chi'_\eta(i\sigma + \nu) = i\chi'(\sigma) + (i\eta/e) + \chi(\nu)$; χ'_η is a character; moreover $\chi', \chi'_1, \ldots, \chi'_{e-1}$ are distinct since their respective values on σ are the distinct elements $\chi'(\sigma), \chi'(\sigma) + 1/e, \ldots, \chi'(\sigma) + (e - 1)/e$ of $\mathbf{Q}/\mathbf{Z}$.

The same result holds for any subgroup $U \leqq G$ (extend by cyclic steps). We conclude that *the order of* $\hat{G}$ *is equal to the order of* G (let $U = \{0\}$ above). *If* $\sigma \in G$, $\sigma \neq 0$, *there exists a character* χ *such that* $\chi(\sigma) \neq 0$ (i.e., $\mathbf{Q}/\mathbf{Z}$ is a *cogenerator* in the category of abelian groups).

64.2 *Bilinear maps into* $\mathbf{Q}/\mathbf{Z}$. Take two additive abelian groups A, B not necessarily finite. Consider a bilinear map $A \times B \to \mathbf{Q}/\mathbf{Z}$, mapping $(a, b) \mapsto a \cdot b \in \mathbf{Q}/\mathbf{Z}$. Connected with this bilinear map are two kernels:

Left kernel $K_L = \{a : a \cdot x = 0 \text{ for all } x \in B\}$
Right kernel $K_R = \{b : x \cdot b = 0 \text{ for all } x \in A\}$.

Assume that the factor group B/K_R is finite. Map $A \to \widehat{B/K_R}$ by $a \mapsto \chi_a$ where $\chi_a(b + K_R) = a \cdot b$ (well defined and is a character). Now what is $\chi_{a_1+a_2}$?

$$\chi_{a_1+a_2}(b + K_R) = (a_1 + a_2)b = \chi_{a_1}(b + K_R) + \chi_{a_2}(b + K_R)$$
$$= (\chi_{a_1} + \chi_{a_2})(b + K_R).$$

Therefore

$$\chi_{a_1+a_2} = \chi_{a_1} + \chi_{a_2}.$$

Hence the map $A \to \widehat{B/K_R}$ is a homomorphism. The kernel of this map is K_L, for, $\chi_a = 0$ means $a \cdot b = 0$ for all b. Therefore $A/K_L \cong$ image. So $A/K_L \to \widehat{B/K_R}$ is a monomorphism. It follows that

$$\text{Order}(A/K_L) \leq \text{Order}(B/K_R)$$

(see 64.1). Similarly if we mapped $B \to \widehat{A/K_L}$ we would get

$$\text{Order}(B/K_R) \leqq \text{Order}(A/K_L).$$

Therefore

$$\text{Order}(B/K_R) = \text{Order}(A/K_L)$$

and

$$A/K_L \cong \widehat{B/K_R}.$$

So the $a \cdot x$ give all characters of B/K_R. We have the following result:

If $A \times B \to \mathbf{Q}/\mathbf{Z}$ *is bilinear and* B/K_R *is finite, then*

$$A/K_L \cong \widehat{B/K_R} \qquad \textit{and} \qquad B/K_R \cong \widehat{A/K_L}.$$

64.3 Let K/k be finite normal with Galois group G abelian of exponent n ($\sigma^n = 1$ for all $\sigma \in G$). Assume

(i) the characteristic p of k does not divide n.
(ii) $X^n - 1$ splits in k

[The assumption (i) implies $X^n - 1$ is separable, for if ε is an nth root of unity

$$\frac{X^n - 1}{X - \varepsilon} = \frac{X^n - \varepsilon^n}{X - \varepsilon} = X^{n-1} + X^{n-2}\varepsilon + \cdots + \varepsilon^{n-1},$$

substitute $X = \varepsilon$ to get $n\varepsilon^{n-1} \neq 0$.]

Let $A = \{\alpha \in K^\times \text{ s.t. } \alpha^n \in k^\times\}$; A is a multiplicative subgroup of $K^\times$. Consider the pairing

$$\chi_{(\ )}(\ )\colon A \times G \to \{\varepsilon\} \qquad \text{(set of } n\text{th roots of unity)}$$

defined by $\chi_{(\alpha)}(\sigma) = \alpha^{\sigma - 1}$. [$\alpha^{\sigma-1} \in \{\varepsilon\}$, for, $(\alpha^{\sigma-1})^n = (\alpha^n)^{\sigma-1} = 1$.] The symbol $\chi_{(\alpha)}(\sigma)$ is linear in α, for, $(\alpha_1\alpha_2)^{\sigma-1} = \alpha_1^{\sigma-1}\cdot\alpha_2^{\sigma-1}$. To show $\chi_{(\alpha)}(\sigma)$ is linear in σ, we observe that $\chi_{(\alpha)}(\sigma) = \alpha^{\sigma-1}$ is a 1-cocycle (in fact it is a 1-coboundary), hence $\chi_{(\alpha)}(\sigma\tau) = \chi_{(\alpha)}(\sigma)\chi_{(\alpha)}(\tau)^\sigma$. But $\chi_{(\alpha)}(\tau) \in k^\times$, so it is fixed under the action of σ. Hence $\chi_{(\alpha)}(\sigma\tau) = \chi_{(\alpha)}(\sigma)\chi_{(\alpha)}(\tau)$. In particular, we have shown $\chi_{(\alpha)}(\ )$ is a character of G.

Let $\chi \in \hat{G}$; then $\chi(\sigma\tau) = \chi(\sigma)\chi(\tau) = \chi(\sigma)\chi(\tau)^\sigma$. Therefore χ is a 1-cocycle. But since $H^1(G, K^\times) = 1$ (63.10), χ is a 1-coboundary. So $\chi(\sigma) = \alpha^{\sigma-1}$ for some $\alpha \in K^\times$. We contend $\alpha \in A$. For, $\chi(\sigma) \in \{\varepsilon\}$, therefore $\chi(\sigma)^n = 1$, so $(\alpha^n)^{\sigma-1} = 1$. This means $\alpha^n \in k^\times$; hence $\alpha \in A$. We have shown *all the characters of G are of the form $\chi_{(\alpha)}(\ )$ for some $\alpha \in A$.*

Now what is K_L? Let $\alpha \in K_L$, then $\chi_{(\alpha)}(\sigma) = \alpha^{\sigma-1} = 1$ for all $\sigma \in G$; or $\alpha^\sigma = \alpha$ for all $\sigma \in G$. Hence $K_L = k^\times$.

What is K_R? Let $\sigma \in K_R$, then $\chi_{(\alpha)}(\sigma) = \alpha^{\sigma-1} = 1$ for all $\alpha \in A$, i.e., $\chi(\sigma) = 1$ for all $\chi \in \hat{G}$. This implies $\sigma = 1$ by the concluding remark in 64.1. Hence $K_R = 1$. Therefore

$$A/k^\times \cong \hat{G} \tag{9}$$

or

$$G \cong \widehat{A/k^\times}.$$

We will next compute the field $k(A)$ (i.e., adjoin to k all the nth roots of elements of k contained in K). We can do this by computing the Galois group U of K over $k(A)$. If $\sigma \in U$, then σ leaves every element of A fixed, i.e., $\sigma \in K_R = 1$. Hence $U = 1$. Therefore

$$k(A) = K.$$

So K is obtained by adjoining to k all $\alpha \in K$ such that $\alpha^n \in k$.

Let $k^\times \subset B \subsetneqq A$. Restrict the bilinear map $\chi_{(\ )}(\ )$ to $B \times G \to \{\varepsilon\}$. The left kernel consists of those $\beta \in B$ such that $\beta^{\sigma-1} = 1$ for all $\sigma \in G$. But this means the left kernel is still $k^\times$. The right kernel is some $U \leqq G$, so

$$G/U \cong B/k^\times.$$

We conclude that the order of G/U is smaller than the order of G, hence $U \neq 1$. Since U leaves the subfield $k(B)$ fixed, we have $k(B) \neq K$.

Conversely, let $k \subset K_0 \subset K$. Since G is abelian, every subgroup is normal, so K_0/k is normal. Moreover K_0 satisfies the same assumptions as K [(i) and (ii) above]. Consequently K_0 is also obtained from some group B, i.e., $K_0 = k(B)$ where $B = \{\beta \in K_0^\times, \text{ s.t. } \beta^n \in k^\times\}$. Thus *there is a one-to-one correspondence between subgroups $B \leqq A$ and the subfields K_0 of K which contain k.* Therefore in this situation we have a substitute for Galois theory (dual of Galois theory, since $A/k^\times \cong \hat{G}$).

64.4 Assume $k^\times$ contains $\{\varepsilon\}$, the nth roots of unity, and $K = k(A)$ where A is a group containing $k^\times$ such that

(i) $A/k^\times$ is finite,
(ii) any $\alpha \in A$ is an nth root of some element of $k^\times$.

Let $\alpha_1, \ldots, \alpha_r$ be representatives of the cosets of $k^\times$ in A, then K is the splitting field of the polynomial

$$(X^n - \alpha_1)(X^n - \alpha_2) \cdots (X^n - \alpha_r).$$

(This follows from the observation that if σ is a k-automorphism of K, then $\sigma\sqrt[n]{\alpha} = \varepsilon_\sigma\sqrt[n]{\alpha}$ where ε_σ is an nth root of unity.) Since $\{\varepsilon\} \subset k$, the automorphisms of K commute on the nth roots $\sqrt[n]{\alpha_j}$, i.e.,

$$\sigma\tau\sqrt[n]{\alpha_j} = \varepsilon_\sigma\varepsilon_\tau\sqrt[n]{\alpha_j} = \varepsilon_\tau\varepsilon_\sigma\sqrt[n]{\alpha_j} = \tau\sigma\sqrt[n]{\alpha_j}.$$

Therefore G is abelian. Moreover for any $\sigma \in G$, σ^n leaves any $\sqrt[n]{\alpha_j}$ fixed, hence G is of exponent n. Summarizing, *under the assumptions made at the start of 64.4, the Galois group of K/k is abelian of exponent n.*

64.5 Returning to 64.3, we now consider the case where the characteristic p of the field k divides the order n of the Galois group of K/k.

We will only consider the special case where $p = n$. Precisely, we are dealing with: k is of characteristic $p \neq 0$. The extension K/k is normal with Galois group cyclic of order p. (No assumptions about roots of unity.) Replace $\mathbf{Q}/\mathbf{Z}$ by $\mathbf{Z}/p\mathbf{Z}$. Let $A = \{\alpha \in K, \text{ s.t. } \alpha^p - \alpha \in k\}$; A is a subgroup of the additive group of K. Indeed if $P\alpha = \alpha^p - \alpha$, then

$$P(\alpha + \beta) = (\alpha + \beta)^p - (\alpha + \beta) = P(\alpha) + P(\beta).$$

Let

$$\chi_{(\ )}(\ ): A \times G \to \mathbf{Z}/p\mathbf{Z}$$

be defined by $\chi_{(\alpha)}(\sigma) = (\sigma - 1)\alpha$. Then $\chi_{(\ )}(\ )$ is bilinear. First we must show $(\sigma - 1)\alpha \in \mathbf{Z}/p\mathbf{Z}$. We know $\alpha^p - \alpha = a \in k$; apply σ to get $(\sigma\alpha)^p - \sigma\alpha = a \in k$. Subtracting the first from the latter we get $P((\sigma - 1)\alpha) = 0$. This implies $(\sigma - 1)\alpha \in \mathbf{Z}/p\mathbf{Z}$, since the kernel of the map which carries $\alpha \in K$ to $P(\alpha) \in K$ is $\mathbf{Z}/p\mathbf{Z}$. [$P(X) = X^p - X$ has at most p roots in K, and $P(0) = 0$, $P(1) = 0$, $P(2) = P(1) + P(1) = 0, \ldots, P(p - 1) = 0$.] Now the bilinearity: $\chi_{(\alpha)}(\sigma)$ is linear in α for

$$(\sigma - 1)(\alpha_1 + \alpha_2) = (\sigma - 1)\alpha_1 + (\sigma - 1)\alpha_2.$$

To show the linearity in σ, observe that $(\sigma - 1)\alpha$ is a 1-coboundary, therefore

$$(\sigma\tau - 1)\alpha = \sigma((\tau - 1)\alpha) + (\sigma - 1)\alpha = (\tau - 1)\alpha + (\sigma - 1)\alpha$$

since $(\tau - 1)\alpha \in k^\times$. In particular we have shown $\chi_{(\alpha)}(\ )$ is a character of G.

Let $\chi \in \hat{G}$, then $\chi(\sigma\tau) = \chi(\sigma) + \chi(\tau) = \chi(\sigma) + \sigma\chi(\tau)$. This shows χ is a 1-cocycle. But since $H^1(G, K^+) = 0$ [in fact $H^r(G, K^+) = 0$ for all $r \in Z$ (15.7)], χ is a 1-coboundary, so $\chi(\sigma) = (\sigma - 1)\alpha$ for some $\alpha \in K$. We contend $\alpha \in A$. For, $P((\sigma - 1)\alpha) = P(\sigma\alpha) - P(\alpha) = 0$, hence σ leaves $P(\alpha)$ fixed. Therefore $P(\alpha) \in k$ and so $\alpha \in A$. We can continue imitating 64.3 to get the left kernel $K_L = k$ and the right kernel $K_R = 1$ and conclude that

$$A/k \cong \hat{G}.$$

Thus, *K is obtained by adjoining to k all $\alpha \in K$ such that $\alpha^p - \alpha \in k$. Moreover there is a one-to-one correspondence between subgroups $B \leqq A$ and subfields K_0 of K which contain k.*

65. Class Formations

65.1 We begin with an example. Let k be a field and K the separable part the algebraic closure of k. (More generally K may be a separable normal extension of k.) Let G be the Galois group of K/k. Let Σ be the set of all finite extensions of k in K. For each $F \in \Sigma$ let G_F be the subgroup of G consisting of those automorphisms which leave F fixed. A topology can be defined on G by letting the family $\{G_F\}_{F\in\Sigma}$ be a fundamental system of neighborhoods of identity. One can prove that G is compact. Then Galois theory literally works for infinite cases provided we consider only closed subgroups of G, i.e., there is a one-to-one correspondence between closed subgroups of G and the subfields of K (see [4, Chapter 6]).

Let $A = K^\times$ be the multiplicative group of K. Then A has a G-module structure. For $F \in \Sigma$ let $A_F = A^{G_F}$ be the submodule of the elements of A which are left fixed under the action of G_F. For E, $F \in \Sigma$, $F \subset E$ if and only if $G_F \geqq G_E$, then $A_F \subset A_E$. Moreover E/F is normal if and only if $G_E \trianglelefteq G_F$, then G_F/G_E is denoted by $G_{E/F}$, and A_E is a $G_{E/F}$-module. Hence we can consider the cohomology groups $H^r(G_{E/F}, A_E)$.

65.2 In local class field theory one deals with the situation described in 65.1 except that k is complete with discrete valuation and a finite residue class field of characteristic p. In global class field theory one is interested in the $G_{E/F}$-module which is called the idèle class group of E for various E. In this case one replaces A by the direct limit of the directed system of monomorphisms of the idèle class groups (see [5, 18, 79]). Thus both in local and global theory one is dealing with the following situation:

A formation $\{G, \{G_F\}_{F\in\Sigma}; A\}$ consists of a group G, an indexed family of subgroups G_F of finite index in G (the indices are called fields), a G-module A, such that

(i) $G_F \subset H \subset G$ implies $H = G_E$ for some $E \in \Sigma$.
(ii) $G_{F_1} \cap G_{F_2} \in \{G_F\}_{F\in\Sigma}$.
(iii) any conjugate of a member of $\{G_F\}_{F\in\Sigma}$ is in $\{G_F\}_{F\in\Sigma}$.
(iv) $\bigcap_{F\in\Sigma} G_F = 1$.
(v) $A = \bigcup_{F\in\Sigma} A^{G_F}$; i.e., every element of A is left fixed by some member of $\{G_F\}_{F\in\Sigma}$.

G is called the *Galois group of the formation; A* is called the *module of the formation.* We say $A_F = A^{G_F}$ is the F *level* of the formation. If $G_E < G_F$, define $E > F$. The pair E, F determine a *layer* (denote it by E/F): A_F is the *ground level*, A_E is the *top level*, and $(G_F: G_E)$ is the *degree* of the layer. If $G_E \trianglelefteq G_F$, then E/F is called a *normal layer*. In this case the factor group $G_F/G_E = G_{E/F}$ acts on A_E and

$$(A_E)^{G_{E/F}} = A_F.$$

Since the degree of each normal layer E/F is finite we can define the cohomology groups $H^r(G_{E/F}, A_E)$. If $E_1/F, \ldots, E_n/F$ are given layers, then there exists a normal layer N/F such that $N > E_i$, $i = 1, \ldots, n$. Namely, let $G_N = \bigcap_{\sigma,i} \sigma G_{E_i}\sigma^{-1}$ where the intersection is over all $\sigma \in G_F$ and all $i \in \{1, \ldots, n\}$. If $D > E > F$, then $G_{D/E} \leqq G_{D/F}$ and both groups operate on A_D, then we can define

$$\text{res}: H^r(G_{D/F}, A_D) \to H^r(G_{D/E}, A_D)$$

and

$$\text{cor}: H^r(G_{D/E}, A_D) \to H^r(G_{D/F}, A_D).$$

Moreover if E/F is a normal layer, then

$$G_{E/F} \cong G_{D/F}/G_{D/E}$$

and for $r > 0$ we can define

$$\text{inf}: H^r(G_{E/F}, A_E) \to H^r(G_{D/F}, A_D).$$

65.3 *Axioms*

AXIOM I $\quad H^1(G_{E/F}, A_E) = 0$ *whenever E/F is a normal layer.*

A formation which satisfies this axiom is called a *field formation.* We have shown in 63.10 that this axiom holds for local class field theory. The verification for global class field theory is more involved (see [5, Chapter 6, Theorem 1]).

If $D > E > F$ and E/F, D/F are normal layers, Axiom I implies the sequence

$$0 \to H^2(G_{E/F}, A_E) \xrightarrow{\text{inf}} H^2(G_{D/F}, A_D) \xrightarrow{\text{res}} H^2(G_{D/E}, A_D)$$

is exact (35.6). If $C > D > E > F$ and E/F, D/F, C/F are normal layers, the diagram of the monomorphisms

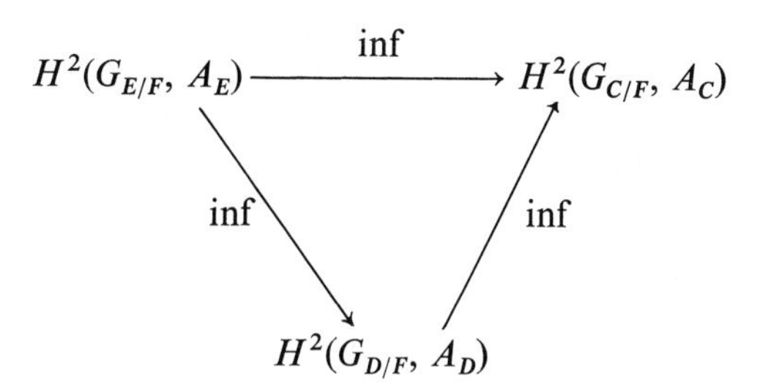

is commutative. Thus we have a directed system of monomorphisms of abelian groups. The direct limit of this system is denoted by

$$H^2(*/F) = \lim_{X>F} H^2(G_{X/F}, A_X), \qquad X \in \Sigma$$

The group $H^2(*/F)$ is called the *Brauer group over F* of the field formation.

AXIOM II *For each field F there is given a monomorphism*

$$\text{inv}_F \colon H^2(*/F) \to \mathbf{Q}/\mathbf{Z}$$

which satisfies the following:

(i) *If E/F is a normal layer of degree n, then* inv_F *maps* $H^2(G_{E/F}, A_E)$ *isomorphically onto the subgroup* $(1/n)\mathbf{Z}/\mathbf{Z}$ *of* $\mathbf{Q}/\mathbf{Z}$.

(ii) *For any layer E/F of degree n,*

$$\text{inv}_E \, \text{res}_{F,E} = n \cdot \text{inv}_F$$

where $\text{res}_{F,E} \colon H^2(*/F) \to H^2(*/E)$ *is induced by the maps*

$$\text{res} \colon H^2(G_{D/F}, A_D) \to H^2(G_{D/E}, A_D).$$

$\text{Inv}_F \, \alpha$ is called the invariant of the element $\alpha \in H^2(*/F)$. Formations which satisfy Axioms I and II are called *class formations.*

The formations of local fields and global idèle classes are class formations. The proofs of these facts are in [5].

One of the main results in class formations is as follows:

The main theorem (Tate). *Let E/F be any normal layer in a class formation,* *then there exists an isomorphism*

$$H^r(G_{E/F}, \mathbf{Z}) \cong H^{r+2}(G_{E/F}, A_E)$$

for all r. (This isomorphism can be constructed explicitly.)

In fact the result obtained in 37.4 was for the purpose of establishing this isomorphism. The proof follows from the verification that Axioms I and II satisfy the hypotheses of 37.4.

Thus in class formations (e.g., local fields or global idèle classes) the cohomology of a normal layer E/F is an invariant of the Galois group $G_{E/F}$ and does not depend on the module A_E.

Bibliography

1. Adamson, I. T. Cohomology theory for non-normal subgroups and non-normal fields. *Proc. Glasgow Math. Assoc.* **2** (1954), 66–76.
2. Alperin, J. L., and Gorenstein, D. The multiplicators of certain simple groups. *Proc. Am. Math. Soc.* **17** (1966), 515–519.
3. Artin, E. *Galois Theory*, Notre Dame Mathematical Lectures (1942).
4. Artin, E. *Algebraic Numbers and Algebraic Functions*, Gordon and Breach, New York, 1967.
5. Artin, E., and Tate, J. *Class Field Theory*, Benjamin, New York, 1967.
6. Atiyah, M. F. Characters and cohomology of finite groups. *Inst. Hautes Etudes Sci., Publ. Math.* **9** (1961), 23–64.
7. Babakhanian, A. Cohomology of finite groups. *Queen's Papers in Pure and Applied Mathematics* **17** (1969).
8. Babakhanian, A. Cohomological triviality and finite p-nilpotent groups. *Arch. Mat.* **21** (1970), 40–42.
9. Baumslag, G., and Gruenberg, K. W. Some reflections on cohomological dimension and freeness. *J. Alg.* **6** (1967), 394–409.
10. Berkson, A. J., and McConnell, A. On inflation–restriction exact sequences in group and Amitsur cohomology. *Trans. AMS* **141** (1969), 403–413.
11. Borevič, Z. I. On homology groups connected with a free group. *Izv. Akad. Nauk SSSR Ser. Mat.* **16** (1952), 365–384.
12. Borevič, Z. I. On homology theory in groups with operators. *Dokl. Akad. Nauk SSSR* **104**, (1955), 5–8.
13. Burgoyne, N., and Fong, P. The Schur multiplicators of the Mathieu groups. *Nagoya Math. J.* **27** (1966), 733–745. Correction, *ibid.*, **31** (1968), 297–304.
14. Cartan, H. Sur la cohomologie des espaces où opère un groupe. *C. R. Acad. Sci., Paris* **226** (1948), 148–150, 303–305.

15. Cartan, H. and Eilenberg, S. *Homological Algebra*, Princeton Univ. Press, Princeton, N. J., 1956.
16. Carton, H., and Leray, J. Relations entre anneaux de cohomologie et groupe de Poincaré. *Colloque Topologie Algebrique, Paris* (1947), 83–85.
17. Cassels, J., and Fröhlich, A. *Algebraic Number Theory*, Academic Press, New York, 1967.
18. Chevalley, C. *Class Field Theory*, Nagoya University, Nagoya, Japon, 1954.
19. Cohn, P. M. Generalisation of a theorem of Magnus. *Proc. London Math. Soc.* **57** (1952), 297–310. Correction, *ibid.*
20. Curtis, C., and Reiner, I. *Representation Theory of Finite Groups and Associative Algebras* Wiley-Interscience, New York, 1962.
21. Eckmann, B. On complexes over a ring and restricted cohomology groups. *Proc. NAS USA* **33** (1947), 275–281.
22. Eckmann, B. Cohomology of groups and transfer. *Ann. Math.* **58** (1953), 481–493.
23. Eilenberg, S., and MacLane, S. Group extensions and homology. *ibid.* **43** (1942), 757–831.
24. Eilenberg, S., and MacLane, S. Cohomology theory in abstract groups, I. *ibid.* **48** (1947), 51–78.
25. Eilenberg, S., and MacLane, S. Cohomology theory in abstract groups, II. *ibid.* **48** (1947), 326–341.
26. Eilenberg, S., and MacLane, S. Cohomology theory of abelian groups and homotopy theory, I. *Proc. NAS USA* **36** (1950), 443–447.
27. Eilenberg, S., and Mac Lane, S. On homology theory of abelian groups. *Can. J. Math.* **7** (1955), 43–55.
28. Eilenberg, S., and Steenrod, N. *Foundations of Algebraic Topology*, Princeton Univ. Press, Princeton, N. J., 1952.
29. Evens, L. The cohomology ring of a finite group. *Trans. AMS* **101** (1961), 224–239.
30. Evens, L. An extension of Tate's theorem on cohomological triviality. *Proc. AMS* **16** (1965), 289–291.
31. Fadeev, D. K. On the theory of homology in groups. *Izv. Akad. Nauk SSSR Ser. Mat.* **16** (1952), 17–22.
32. Fadeev, D. K. On a theorem in the theory of homologies in groups. *Dokl. Akad. Nauk SSSR* **92** (1953), 703–705.
33. Fadeev, D. K. On homology theory for finite groups of operators. *Izv. Akad. Nauk SSSR Ser. Mat.* **19** (1955), 193–200.
34. Gaschütz, W. Kohomologishe Trivialität und äussere Automorphismen. *Math. Z.* **88** (1965), 432–433.
35. Gaschütz, W. Nichtabelsche p-Gruppen besitzen äussere p-Automorphismen. *J. Alg.* **4** (1966), 1–2.
36. Gaschütz, W., Neubüser, J., and Yen, T. Über den Multiplikator von p-Gruppen. *Math. Z.* **100** (1967), 93–96.
37. Golod, E. S. On the cohomology ring of a finite p-group. *Dokl. Akad. Nauk SSSR* **125** (1959), 703–706.
38. Golod, E. S., and Šafarevič, I. R. On the class field tower. *Izv. Akad. Nauk SSSR Ser. Mat.* **28** (1964), 261–272.
39. Gorenstein, D. *Finite Groups*, Harper, New York, 1968.

40. Graham, P. Cohomology of dihedral groups of order $2p$. *Bull. AMS* **72** (1966), 324–325.
41. Green, J. A. On the number of automorphisms of finite group. *Proc. Royal Soc. London, Ser. A.* **237** (1956), 574–581.
42. Gruenberg, K. W. Resolutions by relations. *J. London Math. Soc.* **35** (1960), 481–494.
43. Gruenberg, K. W. A new treatment of group extensions. *Math. Z.* **102** (1967), 340–350.
44. Gruenberg, K. W. Residual properties of infinite soluble groups. *Proc. London Math. Soc.* **7** (1957), 29–62.
45. Gruenberg, K. W. *Cohomological Topics in Group Theory*, Lecture Notes 143, Springer-Verlag, 1970.
46. Grün, O. Über eine Faktorgruppe freier Gruppen, I. *Deutshe Math.* **1** (1937), 772–782.
47. Haimo, F., and Mac Lane, S. The cohomology theory of a pair of groups. *Illinois J. Math.* **5** (1961), 45–60.
48. Hall, P. *Nilpotent Groups*, Lecture Notes, Can. Math. Congr. Summer Seminar (1957).
49. Harrison, D. K. Infinite abelian groups and homological methods. *Ann. Math.* **69** (1959), 366–391.
50. Hochschild, G., and Nakayama, T. Cohomology in class field theory. *ibid.* **55** (1952), 348–366.
51. Hochschild, G., and Serre, J-P. Cohomology of group extensions. *Trans. AMS* **74** (1953), 110–134.
52. Hoechsmann, K. An elementary proof of a lemma by Gaschütz. *Math. Z.* **96** (1967), 214–215.
53. Hoechsmann, K., Roquette, P., and Zassenhaus, H. A. cohomological characterization of finite nilpotent groups. *Arch. Mat.* **19** (1968), 225–244.
54. Hofmann, K. H. Der Schursche Multiplikator topologisher Gruppen. *Math. Z.* **79** (1962), 389–421.
55. Hughes, I. The second cohomology group of one relator groups. *Comm. Pure Appl. Math.* **19** (1966), 299–308.
56. Huppert, B. *Endliche Gruppen*, Springer-Verlag, 1967.
57. Ihara, S., and Yokonuma, T. On the second cohomology groups of finite reflection groups. *J. Fac. Sci. Univ. Tokyo, Sec. I.* **11** (1965), 155–171.
58. Jennings, S. A. The structure of the group ring of a p-group over a modular field. *Trans. AMS* **50** (1941), 175–185.
59. Johnson, D. L. On the cohomology of finite 2-groups. *Invent. Math.* **7** (1969), 159–173.
60. Johnson, D. L. A transfer theorem for the cohomology of a finite group. *ibid.* **7** (1969), 174–182.
61. Johnson, D. L. Non-nilpotent elements of cohomology rings. *Math. Z.* **112** (1969), 364–374.
62. Kawada, Y. Cohomology of group extensions. *J. Fac. Univ. Tokyo, Sec.I* **9** (1963), 471–431.
63. Kochendörffer, R. Über den Multiplikator einer Gruppe. *Math. Z.* **63** (1956), 507–513.

64. Kuo, T. A theorem on the cohomology ring of nilpotent groups. *Proc. AMS* **17** (1966), 1460–1465.
65. Kuo, T. On the exponent of $H^n(G, Z)$. *J. Alg.* **7** (1967), 160–167.
66. Lang, S. *Rapport sur la Cohomologie des Groupes*, Benjamin, New York, 1966.
67. Ledermann, W., and Neumann, B. H. On the order of the automorphism group of a finite group, II. *Proc. Royal Soc. London Ser. A* **235** (1956), 235–246.
68. Losey, G. On dimension subgroups. *Trans. AMS* **97** (1960), 474–486. Correction, *Math. Rev.* **26**, 6260 (1963), 1187.
69. Lyndon, R. C. Cohomology theory of group extensions. *Duke Math. J.* **15** (1948), 271–292.
70. Lyndon, R. C. Cohomology theory of groups with a single defining relation. *Ann. Math.* **52** (1950), 650–665.
71. Massey, W. S. Exact couples in algebraic topology. *ibid.* **56** (1952), 363–396.
72. Magnus, W. Über Beziehungen zwischen höheren Kommutatoren. *J. Reine Angew. Math.* **177** (1937), 105–115.
73. Magnus, W. Beziehungen zwischen Gruppen und Idealen in einem speziellen Ring. *Math. Ann.* **111** (1935), 259–280.
74. Nakayama, T. A theorem on modules of trivial cohomology over a finite group. *Proc. Japan. Acad.* **32** (1956), 373–376.
75. Nakayama, T. A remark on fundamental exact sequences in cohomology of finite groups. *ibid.* **32** (1956), 731–735.
76. Nakayama, T. On modules of trivial cohomology over a finite group. *Illinois J. Math.* **1** (1957), 36–43.
77. Nakayama, T. On modules of trivial cohomology over a finite group, II. *Nagoya Math. J.* **12** (1957), 171–176.
78. Nakayama, T. Cohomology of class field theory and tensor product of modules, I. *Ann. Math.* **65** (1957), 225–267.
79. Neukirch, J. *Klassenkörpertheorie*, B. I. Hochschulskripten 713/713a*, Mannheim, 1969.
80. Neumann, B. H. On some finite groups with trivial multiplicators. *Publ. Math. Debrecen* **4** (1956), 190–194.
81. Northcott, D. *Homological Algebra*, Cambridge, New York, 1962.
82. Ojanguren, M. Algebraischer Beweis zweier Formuln von H. Hopf aus der Homologietheorie der Gruppen. *Math. Z.* **94** (1966), 391–395.
83. Onishi, H. On cohomological triviality. *Proc. AMS* **18** (1967), 1117–1118; On Cohomological Equivalence, *Math Z.* **120** (1971), 221–223.
84. Pareigis, B. Zur Kohomologie endlich erzeugter abelscher Gruppen. *München Akad. Sb.* (1967), 177–193.
85. Parr, J. T. Cohomology of groups of prime square order. *Pacific J. Math.* **17** (1966), 467–473.
86. Passi, I. B. S. Dimension subgroups. *J. Alg.* **9** (1968), 152–182.
87. Quillen, D. The spectrum of an equivariant cohomology ring, I (to appear).
88. Ribes, L. On the cohomology theory of pairs of groups. *Proc. AMS* **21** (1969), 230–234.
89. Ribes, L. Introduction to profinite groups and Galois cohomology. *Queen's Papers in Pure and Applied Mathematics*, No. **24** (1970).
90. Rim, D. S. Modules over finite groups. *Ann. Math.* **69** (1959), 700–712.

91. Rose, J. S. On a splitting theorem of Gaschütz. *Proc. Edinburgh Math. Soc.* **15** (1966), 57–60.
92. Scott, W. R. *Group Theory*, Prentice-Hall, Englewood Cliffs, N. J., 1964.
93. Serre, J-P. *Corps Locaux*, Hermann, Paris, 1962.
94. Serre, J-P. *Cohomologie Galoisienne*, Springer Notes, 1964.
95. Snapper, E. Cohomology of permutation representations, I. *J. Math. Mech.* **13** (1964), 133–161.
96. Snapper, E. Cohomology of permutation representations, II. *ibid.* **13** (1964), 1047–1064.
97. Snapper, E. Duality in the cohomology ring of transitive permutation representations. *ibid.* **14** (1965), 323–336.
98. Snapper, E. Spectral sequences and Frobenius groups. *Trans. AMS* **114** (1965), 113–146.
99. Snapper, E. Inflation and deflation for all dimensions. *Pacific J. Math.* **15** (1965), 1061–1081.
100. Stallings, J. Homology and central series of groups. *J. Alg.* **2** (1965), 170–181.
101. Stallings, J. On torsion-free gree groups with infinitely many ends. *Ann. Math.* **88** (1968), 312–334.
102. Stammbach, U. Anwendungen der Homologietheorie der Gruppen auf Zentralreihen und auf Invarianten von Präsentierungen. *Math. Z.* **94** (1966), 157–177.
103. Stammbach, U. Über freie Untergruppen gegebener Gruppen. *Comment. Math. Helv.* **43** (1968), 132–136.
104. Swan, R. G. The *p*-period of a finite group. *Illinois J. Math.* **4** (1960), 341–346.
105. Swan, R. G. Minimal resolutions for finite groups. *Topology* **4** (1965), 193–208.
106. Tate, J. The higher dimensional cohomology groups of class field theory. *Ann. Math.* **56** (1952), 294–297.
107. Tate, J. Nilpotent quotient groups. *Topology* **3** (1964), 109–111.
108. Venkov, B. B. Cohomology algebras for some classifying spaces. *Dokl. Akad. Nauk SSSR* **127** (1959), 943–944.
109. Vermani, L. R. An exact sequence and a theorem of Gaschütz, Neubüser and Yen on the multiplicator. *J. London Math. Soc.* **1** (1969), 95–100.
110. Wall, C. T. C. On the cohomology of certain groups. *Proc. Cambridge Phil. Soc.* **57** (1961), 731–733.
111. Weiss, E. A deflation map. *J. Math. Mech.* **8** (1959), 309–329.
112. Weiss, E. *Cohomology of Groups*, Academic Press, New York, 1969.
113. Wiegold, J. Multiplicators and groups with finite central factor groups. *Math. Z.* **89** (1965), 345–347.
114. Wong, W. J. A cohomological characterization of finite nilpotent groups. *Proc. AMS* **19** (1968), 689–691.
115. Yang, K. On some finite groups and their cohomology. *Pacific J. Math.* **14** (1964), 735–740.
116. Yokonuma, T. On the second cohomology groups of infinite discrete reflection groups. *J. Fac. Scil Univ. Tokyo, Sec. I* **11** (1965), 173–186.
117. Zassenhaus, H. *The Theory of Groups*, Chelsea, New York, 1956.

Notation Index

Subject Index